Volunteer Tourism and the Moral Self

Volunteer Tourism and the Moral Self, offers a new lens to conceptualise volunteer tourism through the 'moral self'. It moves the conceptualisation of volunteer tourism to the broader discussion around ways of being and becoming a moral self. It is the first volume of ethnographic research of Asian experiences of volunteer tourism which has been a field of study premised on Western participants and weighted with Western assumptions and ethical models.

Drawing on concepts and theories in geography, anthropology, sociology, tourism and education, *Volunteer Tourism and the Moral Self* explores how a moral self is cultivated, experienced and (hopefully) re-invented through volunteer tourism. It navigates with volunteer tourists from Hong Kong and Taiwan to examine how volunteer tourism has become a social trend. This social trend emerges from the interplay of institutionalised service obligation in schools and the culturally rooted ethical dispositions. It also manifests the search for rebuilding social ties in different forms of moral communities and new ways of being.

Yim Ming Connie Kwong is a human geographer, with an interdisciplinary background in cultural geography, tourism and sustainability. She holds a PhD in Geography from Durham University and an MPhil in Geography and BSocSc. (Geography and Sociology) from the University of Hong Kong. Her research interests lie in three interdisciplinary areas: 1) cultures, values, identities and practices; 2) moral geographies, tourism and development and 3) community building and sustainable co-development. She is currently a postdoctoral researcher in Leibniz Centre for Tropical Marine Research (ZMT). She has conducted interdisciplinary and transdisciplinary research and fieldwork as well as supervising student projects and theses of different levels in East, Southeast and Central Asia, Latin America and East Africa. She is co-editor of a recent volume 'Navigating the Field: Postgraduate Experiences in Social Research'.

Routledge Advances in Tourism and Anthropology: People, Place and World

Series Editors:

Dr Catherine Palmer (University of Brighton, UK) C.Palmer3@brighton.ac.uk
Dr Jo-Anne Lester (University of Brighton, UK) J.Lester@brighton.ac.uk

To discuss any ideas for the series please contact Faye Leerink, Commissioning Editor: faye.leerink@tandf.co.uk or the Series Editors.

This series draws inspiration from anthropology's overarching aim to explore and better understand the human condition in all its fascinating diversity. It seeks to expand the intellectual landscape of anthropology and tourism in relation to how we understand the experience of being human, providing critical inquiry into the spaces, places, and lives in which tourism unfolds. Contributions to the series will consider how such spaces are embodied, imagined, constructed, experienced, memorialized and contested. The series provides a forum for cutting-edge research and innovative thinking from tourism, anthropology, and related disciplines such as philosophy, history, sociology, geography, cultural studies, architecture, the arts, and feminist studies.

Tourism and Embodiment
Edited by Catherine Palmer and Hazel Andrews

Tourism Encounters and Imaginaries: The Front and Back Stage of Tourism Performance
Edited by Frances Julia Riemer

Folklore, People and Place
International Perspectives on Tourism and Tradition in Storied Places
Edited by Jack Hunter and Rachael Ironside

Volunteer Tourism and the Moral Self
Ethnographic Research of Non-Western Tourists
Yim Ming Connie Kwong

For more information about this series please visit: www.routledge.com/Routledge-Advances-in-Tourism-and-Anthropology/book-series/RATA

Volunteer Tourism and the Moral Self

Ethnographic Research of Non-Western Tourists

Yim Ming Connie Kwong

LONDON AND NEW YORK

First published 2025
by Routledge
4 Park Square, Milton Park, Abingdon, Oxon OX14 4RN

and by Routledge
605 Third Avenue, New York, NY 10158

Routledge is an imprint of the Taylor & Francis Group, an informa business

British Library Cataloguing-in-Publication Data
A catalogue record for this book is available from the British Library

ISBN: 978-1-032-53843-3 (hbk)
ISBN: 978-1-032-53844-0 (pbk)
ISBN: 978-1-003-41392-9 (ebk)

DOI: 10.4324/9781003413929

Typeset in Times New Roman
by SPi Technologies India Pvt Ltd (Straive)

Contents

Acknowledgements

This book is one of the outcomes of my doctoral research at the Department of Geography at Durham University. My PhD topic was developed based on discussions with my postgraduate colleagues about community development projects, experiences of participating in and organising voluntary services and service trips as well as observation of the volunteering culture on the university campus. The research project was then further developed with the inspirations and mentoring of Professor Mike Crang and Professor Simone Abram at Durham University. They gave me a tremendous amount of guidance, support and encouragement to accomplish the project as well as other endeavours. Along the way, my family has also given me endless support, care and understanding.

The analysis and arguments developed in this book draw primarily on my ethnographic research conducted with volunteer tourists from Hong Kong and Taiwan who participated in service trips to Cambodia. It is then supported by my observations and reflections from previous volunteering experiences and more recent ethnographic research in other contexts. I would like to thank all those who were involved in and supported my fieldwork, especially my research informants who were so open in sharing their experiences and thoughts. I am very glad that they are not just my research informants; we have become friends. I would also like to thank all the colleagues and friends with whom I have spoken and who have inspired new ideas for this book.

I would also like to express my gratitude to various sources of funding throughout the course of my PhD studies. Without the postgraduate scholarship from the Sir Robert Black Trust Fund in Hong Kong, the True Light Scholarship from my secondary school True Light Girls' College, the Global Citizenship Programme Scholarship from Ustinov College in Durham, the postgraduate stipend from the Department of Geography in Durham, and the PhD grant from the Allan and Nesta Ferguson Charitable Trust, I could not have completed the research project. The Norman Richardson Postgraduate Research Fund also provided additional support for my fieldwork.

I am very grateful for all the support that I received throughout the preparation and production of this book, including the advice and encouragement of the Series Co-Editors Dr. Catherine Palmer and Dr. Jo-Anne Lester, the feedback of the anonymous reviewers, and the support of the editorial team at Routledge and Straive (Faye Leerink, Prachi Priyanka, Akansha Bharali, Madii Cherry-Moreton, Jayapriya Thillaikkarasu, Copy-editors and Typesetters of Straive).

1 Introduction

Situating volunteer tourism and the moral self in Asia

Volunteer tourism in Asia

In 2012, I attended the launch ceremony of the Gallant Ho Experiential Learning Centre in the University of Hong Kong (HKU). The centre was established to promote service learning with a generous donation by Dr. Gallant Ho, a supporter of public service. This establishment, together with the HKU SERVICE 100 Fund, supports HKU students to engage in local and overseas voluntary services. Walking along the corridors or podium of HKU's campus, one cannot miss the walls of posters or booths set up by the student organisations inviting participation in service trips. This is a common sight on the walls across universities in Hong Kong: posters promoting service-learning trips with the promise of a meaningful and lifelong inspirational experience (Cheung, 2016).

The rise of volunteerism and overseas volunteering expeditions in Asia is broadly contemporaneous with that in the West. Organisations such as the Voluntary Service Organisation and British Trust for Conservation Volunteers in the United Kingdom were established in the late 1950s and the Peace Corps in the United States in the early 1960s to promote voluntary service (Wearing & McGehee, 2013). The Japan International Cooperation Agency established the Japan Overseas Cooperation Volunteers in 1965 following the spirit of Peace Corps in the United States (TaiwanICDF, 2004). In Hong Kong, the Agency for Volunteer Service was founded in 1970 to promote and develop volunteerism to build a civil society and caring community, in the meantime there is a long history of wealthy families in Hong Kong setting up philanthropic foundations which support poverty alleviation and education (Cheung, 2016). Korea International Cooperation Agency, established in 1991 to improve relations with the international community through developmental projects and university volunteering, has played a significant role in supporting this governmental effort (Lee & Yen, 2015). In Singapore, the rise of volunteer tourism was driven by two developments: the compulsory community involvement programme implemented by the Ministry of Education in Singapore for all pre-tertiary schools in 1997 and the introduction of the Youth Expedition Project in 2000 under the Singapore International Foundation (Sin, 2009).

DOI: 10.4324/9781003413929-1

While Asian individuals are actively volunteering, there is still a paucity of studies on the trend and development of volunteer tourism in Asia. Yea (2018) has argued that the Asia-Pacific region has grown as a significant contributor to the rest of the world through international volunteering rather than being simply a recipient of international volunteers and aid. However, volunteer tourism has long been conceived as a Western-based pilgrimage, with volunteers from Western societies in the Global North sending aid to the developing countries in the Global South. The existing academic literature on volunteer tourism has been very largely premised on Western participants, weighted with Western assumptions and ethical models (see also Lo & Lee, 2011; Pan, 2017; Polus, Carr, & Walters, 2023; Qi, 2020; Wang & Brown, 2024). Voices from Asian volunteer tourists have largely been underrepresented, except in limited studies on volunteer tourists from Hong Kong (Lo & Lee, 2011), Singapore (Chen, 2018; Sin, 2009; Yea, 2018), South Korea (Lee & Yen, 2015) and Taiwan (Pan, 2012, 2017; Tu, 2011), and Chinese volunteering within China (Wang & Brown, 2024). Most focus on the motivations of the volunteers and impacts on them, except Chen (2018) who looks more deeply into how interpersonal relationships in international volunteering affect development impact through a relational approach, and Yea (2018) who relates the motivations and impacts to wider social and cultural issues in the case of domestic forms of development volunteering for migrant rights within Singapore.

Volunteer Tourism and the Moral Self widens the discussion to a non-Western context, problematising the existing academic literature on volunteer tourism. With a focus on Asia which is a significant but neglected origin of volunteer tourists, it responds to the urge brought up by Porananond and King (2014) for 'Asianising the field' through more research on the emerging phenomenon of Asian tourists/tourism within Asia, and a more recent call by Sin, Mostafanezhad, and Cheer (2021) for 'radical shifts towards more inclusive epistemology and praxis' given the 'recentering of tourism geographies in the Asian Century'.

Volunteer tourism and the 'self'

Named by booking.com as the number-one travel trend in 2019, the number of conscious individuals taking part in volunteer tourism programmes has grown from 1.6 million annually in 1990 to around 10 million in more recent years, spending one to three billion dollars on these trips per year (CBI, 2020; Gharib, 2021). Wearing (2001) defines volunteer tourists as

> those tourists who, for various reasons, volunteer in an organized way to undertake holidays that might involve aiding or alleviating the material poverty of some groups in society, the restoration of certain environments or research into aspects of society or environment.
>
> (p. 1)

He also regards volunteer tourism 'as a development strategy leading to sustainable development and centering the convergence of natural resource qualities, locals and the visitors that all benefit from tourism activity' (p. 12). Central to this definition is that volunteer tourism brings positive impacts to the environment or host communities as well as intrinsic rewards or positive changes in volunteer tourists. Volunteers usually take part in short-term programmes organised by 'sending organisations' which include not-for-profit organisations, social enterprises, private companies, universities, religious organisations, conservation agencies, and charities (Raymond & Hall, 2008).

The research foci of volunteer tourism have changed over the past two or three decades. The focus of earlier studies was primarily on pre-trip motivations,[1] impacts and outcomes.[2] Studies later have shifted to examine power relations and exploitation[3] as well as volunteer tourism as a form of colonialism and neo-colonialism that deepens dependency.[4] There are also lines of research extensively on the geography of compassion, care and responsibility in volunteer tourism that could be perceived as a development tool,[5] and the neoliberal moral economy.[6] Within the hopeful tourism scholarship, a line of inquiries has emerged to explore the transformative potential of volunteer tourism and self-development.[7]

The 'moral turn' in geography in the 1990s (Smith, 1997) and that in tourism studies (Caton, 2012) has brought up three themes that I found especially significant for this book, naming responsibility, obligation and care (Hall, 2011). As volunteer tourism has been conceived as a form of responsible tourism or neoliberal governance, the rhetoric of self-making work through caring for the distant others and enacting global responsibility has successfully asserted responsibility into the choice of travel, a travel experience injected with the moral responsibility of work (Sin, Oakes, & Mostafanezhad, 2015). The 'gap year' for British students or 'The Big OE' for Australasians is no longer about irresponsible 'dropping out'; it is more about travelling as a global citizen, contributing to community development and the wellbeing of local communities through volunteer tourism (Lyons, Hanley, Wearing, & Neil, 2012). Relevant terms like 'care', 'responsibility' and 'awareness' have become prevalent in the advocacy for a more self-consciously moral choice of tourism (Butcher, 2003, 2014).

Despite the expansive body of literature on volunteer tourism, there is no particular work to situate the conceptualisation of this form of tourism around the 'self', to unpack what the 'self' means and entails in the self-making through volunteer tourism. *Volunteer Tourism and the Moral Self* offers an in-depth discussion and conceptualisation of volunteer tourism as a social trend, to understand what frames volunteer tourism as an appropriate choice for crafting a class of moral self, what the moral self seeks, how the moral self is developed and what this cultivation of the moral self brings to the self and beyond. It invites discussion on what and how responsibility, obligations and ethics of care are negotiated and intersected, and then embodied through action and practice. With a focus on non-Western tourists, specifically Asians, it explores the cultural particularities in these dimensions of conceptualisation.

The moral self

Caton (2012) defines morality as the 'human imaginative and discursive capacity for considering how things should be, as opposed to describing how things are – what is sometimes referred to as the "is" versus "ought" distinction' (p. 1907). Stafford (2013) posits that ethics is instantiated in agency whereas morality is instantiated in structure. Mostafanezhad and Hannam (2014) summarise the difference between morality and ethics as that ethics refers to the codes of behaviour for a social group while morality is a broad belief system 'regarding how things ought to be across the range of human experience' (p. 4). Smith (1997), however, has argued that definitional disputes offer few gains and that

> nothing much is lost if both terms are taken to mean the same: having to do with evaluation of human conduct, with what is right or wrong or good or bad, with what people ought or ought not to do, and with the quality of their actions or characters, in contexts which are not merely matters of etiquette or prudence.
>
> (p. 584)

In this book, 'moral' is used in the 'moral self' as the primary term to tie the discussion to the themes of care, responsibility and obligation that are significant in the 'moral turn', while moral/morality and ethical/ethics are used interchangeably in some contexts.

Both morality and ethics have been perceived as very philosophical, implicit and exceptional. However, it is believed that some ethical situations and practices are ordinary and routine (Lambek, 2010a; Stafford, 2013). Singer (2013) argues that

> it is vital that ethics not be treated as something remote… We cannot avoid involvement in ethics, for what we do – and what we don't do – is always a possible subject of ethical evaluation. Anyone who thinks about what he or she ought to do is consciously or unconsciously, involved in ethics.
>
> (p. v)

As Lambek (2010a) suggests, ethical life is therefore observable not only in exceptional circumstances such as 'moral dilemmas', but also in routine and everyday ones. Ethics is intrinsic to speech and action in everyday circumstances as well as cultural values. It is not just a component of an action but also an action (Lambek, 2010a, 2010b), 'the ordinary is intrinsically ethical and ethics intrinsically ordinary' (Lambek, 2010a, p. 3). This broadly follows Bourdieu's ideas in accentuating how ethical judgements depend on tacit understandings of how things are and happen as unconscious responses rather than explicit, conscious reasoning and reflections (Lambek, 2010a). 'Ordinary', as Lambek (2010a) perceives, implies an ethics that occurs with undue attention, grounded in agreement and norms rather than rules and regulations, in practice rather than knowledge or belief. In other words, routine ethical circumstances are often dealt with via tacit understandings and micro-processes of everyday life rather than via explicit philosophising, theory building or decision-making. Responding to this approach, Stafford (2013) emphasises and posits that ethics is also the subject of explicit and conscious deliberation which is a routine dimension of action, and thus 'explicit ethical reflection, explicit discussion of ethical matters and explicit judgement and decision-making in relation to ethical demands are also "ordinary" aspects of human life' (p. 5). This ordinariness, implicit or explicit, implies a 'thick morality' (Walzer, 1994), which is local, contextual and culturally particular. Building on these ideas, ordinary ethics provides a foundation of analysis in this book, guiding us not to focus on Ethics with capital E but on the ordinary ethics or the local, everyday thick morality.

There is an expanding literature discussing ordinary ethics in relation to ethical consumption as consumption provides a platform for expressing personal ethical beliefs and exercising ethical motivations (Hall, 2011; Wilk, 2001). Cloke, Johnsen, and May (2007) see ordinary ethics as 'complex everyday caring and relations with others which are widespread through society' (p. 1099). Such circumstance is embedded in routines and deeply ingrained habits (Barnett, Cloke, Clarke, & Malpass, 2005; Barnett, Cloke, Clarke, & Malpass, 2010b). Daya (2016) has looked into how people's ethical beliefs and recognition of a shared identity and others are embodied through buying handmade crafts in Cape Town, and revealed that ethical relationships between consumers and producers are central to the business of trade in informal and market spaces in the sense that ordinary ethics of care for others go beyond explicit, rational responsibility to cultivate oneself. Given the disposition of ordinary ethics, the question moves forward to how such disposition is 'worked up' (Barnett et al., 2005; Popke, 2006), how motivations are translated into actual practices and how these practices are then morally made sense of from the embedding of ethics in practices. This also brings up the question of how a choice becomes appropriate or ethical.

As ethics is intrinsic to everyday circumstances in different cultures, cultural specificity has to be taken into account. Such practice may arise from cultural norms that an individual encounters and negotiates in daily life and ethics are operationalised by the individuals differently (Subrahmanyan, Stinerock, & Banbury, 2015), or embedded in a cultural context of global consumer capitalism (Varul, 2009). Crang and Hughes (2015) pose the question: 'What happens when the consumers may be neither so distant, nor so affluent?' (p.131), leading us to think about the thick morality in non-Western contexts. Putting this in a tourism context, ordinary ethics provides a useful lens in unpacking practices, encounters and experiences. However, it has not yet been explicitly deployed to the study of volunteer tourism.

In the following, I am going to delineate the theoretical framework that helps to examine the meaning, constitution and development of the moral self through (1) conscious self-formation, (2) unconscious inculcation of dispositions, (3) quality or value accumulation and (4) the moral self and the others.

Conscious self-formation

Foucault's later work on ethics, subjectivation and technologies of the self has provided a useful lens in unpacking the development of this moral self. Subjectivation, the relation of the self to the self, brings up discussions on how human beings constitute themselves as a subject and relate morally to themselves and others. For an action to be moral, it should not be reduced to only conforming to rules and laws; it should involve a relationship with the self, a self-formation of ethical subject to get prepared for 'a certain complete achievement of life' (Foucault, 1988, p. 31). The moral goal of forming part of the self as the object of their own moral practice and deciding on the mode of being requires the individual to act upon oneself, to improve and transform oneself. According to Foucault (1985), there is 'no moral conduct that does not call for the forming of oneself as an ethical subject; and no forming of the ethical subject without "modes of subjectivation" and an "ascetics" or "practices of the self" that support them' (p. 28). Therefore, understanding ways in which the individual is urged to constitute the self as an ethical subject requires the study of the forms of moral subjectivation and the practices of the self for the transformation (Foucault, 1985). This self-forming work and practices of the self emphasise the forms of relations with the self as well as methods and techniques to enable the transformation into a certain mode of being. This form of relations with the self, 'care for the self' in Foucault's term, results in dispositions of ethics (Fornet-Betancourt, Becker, Gomez-Müller, & Gauthier, 1987). Different forms of care produce different forms of the self. Thus, Foucault suggests us to find our own applications and in this endeavour actively shape our ethos through an aesthetics of self-development in order to become ethical subjects of our conduct (Feighery, 2011; Foucault, 1985). Foucault's concept of subjectivation and care for the self helps us understand the kind of self-fashioning ethics of care and the technologies to transform the self in the contemporary society.

Such self-forming work can be understood with Foucault's concept of technologies of the self. It refers to the means to govern 'the conduct of individuals and submit them to certain ends' (Foucault, 1988, p. 18). These means

> permit individuals to effect by their own means or with the help of others a certain number of operations on their own bodies and souls, thoughts, conduct, and way of being, so as to transform themselves in order to attain a certain state of happiness, purity, wisdom, perfection, or immortality.
>
> (Foucault, 1988, p. 18)

These practices focus on the 'hermeneutics of the self, rather than the more practical business of the actual training of individuals in particular social activities' (Burkitt, 2002, p. 221). These modes of training and modification are not just about acquiring certain sets of skills but also certain attitudes, resulting in an individual with the dispositions and capacities to act morally (Burkitt, 2002; Foucault, 1988). Technologies could be understood as practical rationality that informs and accompanies practical action which aims to instil habitus, skills and capacities and, later, to provide the reflexive powers to reason about their virtues in the self-formation process (Burkitt, 2002; more details on 'habitus' in the next section). Such practical rationality guides us to take appropriate action in complex social contexts, in which 'appropriate' represents a notion of balanced judgement based on moral concerns (Sanderson, 2006). These techniques are also traditional in the sense that they are cultural and passed down through social groups, or ingrained in social relations (Burkitt, 2002).

The means to govern the conduct of individuals and submit them to certain ends leads us to the discussion of the neoliberal discourses of responsibilisation. Under neoliberalism, agenda, burden or efforts of resolving social problems, which were primarily state's business, have been shifted to the individuals, resulting in them taking up the responsibility. Individuals are positioned by the governmental rationalities that they need to work on themselves and their conduct in order to transform themselves (Barnett, Cloke, Clarke, & Malpass, 2010a). Responsibilisation implies that individuals must be actively responsible for carrying out these activities and their outcomes (Burchell, 1993; Kelly, 2001). It also encourages individuals to be self-enterprising to 'conduct themselves in accordance with the appropriate (or approved) model of action' (Burchell, 1993, p. 276). Shamir (2008) also sees responsibilisation as a call for a response or action in particular, 'an interpellation which constructs and assumes a moral agency and certain dispositions to social action that necessarily follow' (p. 4). Living in an age of responsibilisation, Shamir (2008) further explains responsibilisation as a technique of governance,

> fundamentally premised on the construction of moral agency as the necessary ontological condition for ensuring an entrepreneurial disposition in the case of individuals and socio-moral authority in the case of institutions. Neoliberal responsibilisation is unique in that it assumes a moral agency which is congruent with the attributed tendencies of economic-rational actors: autonomous, self-determined and self-sustaining subjects.
>
> (p. 7)

As responsibilisation has reconfigured the roles and identities of individual actors, these subjects have to be 'made up' through the mobilisation of techniques to actively undertake and perform self-governing tasks (Beck, 1992; Kelly, 2001). The governmental rationalities 'elicit, promote, facilitate, foster, and attribute various capacities, qualities and statues to particular agents' (Dean, 1999, p. 32; as cited in Barnett et al., 2010b, p. 44). Such rhetoric has led to 'deepening of personal responsibility' (Patrick, 2012; as cited in Garthwaite, 2017) and these agents are in need of empowerment while being actively responsibilised (Garthwaite, 2017).

The process of self-formation to become an ethical subject can also be conceptualised through the idea of moral selving. As mentioned earlier, volunteering involves both improving oneself as well as helping others, and some researchers have posited that it is hard to distinguish clearly between altruism and self-interest and thus it is reasonable to possess both (Tomazos & Butler, 2010; Varul, 2009). Allahyari (2000) introduces 'moral selving' to refer to the 'work of creating oneself as a more virtuous, and often more spiritual, person [...] one type of deeply emotional self work' (p. 4). Moral selving is the process of shaping, striving, creating, building, and sculpting the moral self – 'the cognitive, and emotive, self-related elements of the social self' in the form of a disposition of moral responsibility (Allahyari, 2000, p. 4).

Moral selving can also be understood as the mediated work of crafting a moral self through practices that acknowledge responsibilities to others, either explicit display or implicit, humble mode of conduct (Barnett et al., 2005). Using a case study of fairtrade in the United Kingdom, Barnett et al. (2005) argue that the ordinarily ethical everyday consumption routines are performances and mediated practices of constructing a life through negotiating consumer choices about personal conduct, and such ethical expression and exercise involves governing both the consumption and the consuming self. These everyday practical dispositions are re-articulated with campaigns and politics that register individual responsibility into broader forms of collective accountability (Barnett et al., 2005).

Boluk (2011) investigates the notion of moral selving in the context of ethical travel using the case study of Western tourists going on holiday certified as Fair Trade Tourism in South Africa. Fair Trade Tourism, a kind of pro-poor tourism or ethical consumption, can be conceived of conspicuous consumption as the consumption in itself is hedonistic, especially when the tourists use it as a means to demonstrate their ethical self, or their 'elite' status from their education and perceived higher level of awareness (Boluk, 2011). Participation in Fair Trade Tourism allows the tourists to appear as if they were more consistent in their motivations and demonstrated self-image because their ethical concerns seem to have influenced their travel behaviour, although the languages they used sometimes contrasted their demonstrated virtuous tendencies and aspirations (Boluk, 2011). Hence, the consumption of these ethical tourism products or similar ethical activities represents the power of the consumers to create and choose the self, and to assist themselves in self-fashioning as ethical, concerned, educated and good human beings (Boluk, 2011; Slater, 1997).

Similar self-fashioning work communicated through volunteer tourism has been seen in a relatively cynical term. Participation in volunteer tourism has received criticism over the debate between altruism and self-interest, having more positive impacts on the volunteers rather than the host community, or failure to live up to its promise as a vehicle for making positive social and environmental changes to the wider society (Coghlan & Gooch, 2011). McGehee (2014) has noticed the critique of volunteer tourism as 'a potential purveyor of western neoliberalism and neo-colonial notions of "us versus them"' (p. 850). Sin (2009) found that 'many volunteer tourists are typically more interested in fulfilling objectives relating to the "self"' (p. 497). Butcher and Smith (2010) have also analysed that 'the "desire to make a difference"...[has become] connected to lifestyle... [and] closely linked to a narrative of personal growth' (p. 33). These discussions then critique altruism as the claimed foundational motivations and purpose of volunteer tourism. Ateljevic (2009) argues that the academic literature has demonstrated a more aggressive trend of criticising rather than critiquing volunteer tourism. She asserts that the various issues and problems facing human beings have been leading people to shift from the current unsustainable, materialist paradigm to the transmodernity paradigm to 'search for a new worldview of higher values and more responsible, meaningful lives' (Ateljevic & Tomljenovic, 2016, p. 26). A 'hopeful tourism' scholarship with a 'values-based, unfolding transformative perspective' is believed to be needed to offer active hope which helps to visualise ideas and formulate corresponding plans to accomplish its ends (Pritchard, Morgan, & Ateljevic, 2011, p. 942). Ateljevic and Tomljenovic (2016) propose, in order to be transformational, 'tourism must stimulate change by provoking a deep questioning of the purpose and meaning of people's life through empathic, engaged, authentic and invited, rather than imposed, encounters with the lives of others' (p. 30). Following this, a growing need for transformative and conscious travel to find the means to change one's life(style) and impacts on the destinations has been seen (Ateljevic & Tomljenovic, 2016). Correspondingly, rather than cynically criticising volunteer tourism for focusing on personal development of volunteers and not bringing sustainable impacts to the host community or local environment, we have started to see views on transformative potential of volunteer tourism. This gives rise to the idea of transformation of the self by better understanding the self and becoming more conscious through travelling experience.

Lepp (2008) examined how volunteers have been influenced by the voluntary work in Kenya and found that volunteers 'developed a new perspective on life at home and discovered an intrinsic need for meanings and purposes of their lives' (p. 98). The findings of the study by Lo and Lee (2011) on Hong Kong volunteer tourists indicated changed views of life, personal development and growth, and influence on life directions. Taiwanese volunteer tourists have reported similar changes in view which resulted in learning to appreciate a slower life pace and self-relaxation at the same time changing choice of travel destination (Pan, 2017). Nonetheless, these previous studies primarily focus on listing out reported outcomes without much looking into the transformative process or underpinning that contributes to the making and (new) experience of the moral self.

Some studies have attempted to delve into the transformative potential through more conceptual understanding of the changes in worldview or self-identity, unravelling how volunteer tourism experience has led to further participation and/or (re)working of the volunteer tourists. McGehee and

Santos (2005) examined how the experience of volunteer tourists informed their involvement in social movements, by drawing on theoretical perspectives of social psychology and resource mobilisation. The findings indicated that participation in volunteer tourism has positive impacts on intended post-trip social movement activities and support for social activism from the establishment and expansion of social networks and raised consciousness (McGehee & Santos, 2005). Although not taking part in the same programme, volunteer tourists of their study continued participation in movements of similar cause or adopted different ethical practices. Kontogeorgopoulos (2017) explored the importance of existential authenticity in motivations of and activities undertaken by volunteer tourists travelling to Thailand and found that a volunteer tourism experience is driven by a desire for existential authenticity which facilitates the self-making. Zahra (2010) found through a longitudinal study how volunteer tourists experienced a change in life course. When the encounter of poverty and suffering intersected with the cheerfulness of the local communities amid lack of resources, volunteers were found to experience a cathartic and life-changing experience which showed impacts on their everyday life even eight years after participation. Zahra (2010) has concluded that volunteer tourism has the potential of transforming and reworking the self which corresponds to what Wearing (2001) suggests that volunteer tourism causes 'value change and changed consciousness in the individual that will subsequently influence their lifestyle' (p. x.).

Transformative learning has received growing attention and interest in studying volunteer tourism. Mezirow's (1991) transformative learning theory is considered to be a useful framework for viewing volunteer tourists' experience, while efforts are required by the sending organisations to help participants 'developing a plan of action' and 'trying out new roles and gaining feedback' to complete the transformative process (Coghlan & Gooch, 2011). By doing so, it is more likely to move volunteer tourism beyond the rhetoric 'doing something worthwhile' or 'making a difference' to life-changing in the long run that will benefit both volunteers and the volunteered (Coghlan & Gooch, 2011). Hammersley (2014) has extended the conversation by arguing the importance of structural opportunities for both pre-trip preparation and post-trip debriefing of volunteers in facilitating transformation to take place, during which expedition leaders would play a key role in providing support through the process of disorienting dilemma or effecting changes in the frames of reference (Taylor, 2008). Tomazos and Murdy (2023) posit that the individualised transformational experiences may result in a perpetual liminality in which the volunteers are trapped in limbo. This implies a significant role played by the sending organisations in their marketing as well as the meaningful integration of such experience in the programme. However, only a few studies have pinpointed the importance to guide volunteer tourists to seek greater understanding and knowledge prior to the trip and the need to incorporate critical reflection of their experiences upon return (e.g. Grusky, 2000; Raymond, 2008; Raymond & Hall, 2008; Simpson, 2004, 2005).

Literature with a focus on post-trip reflection or learning is still scant (Henry, 2019; see also Coghlan & Weiler, 2018; Couch & Georgeou, 2017; Ong, King, Lockstone-Binney, & Junek, 2018). More research is needed on the transformative learning process and its potential positive impacts to develop the moral self. This book attends to this need to discuss its potential in developing the moral self from the conceptualisation of volunteer tourism as an appropriate choice for conscious self-formation.

Unconscious inculcation of dispositions

Looking at the process of instilling dispositions that inform an individual's practices and actions, it allows us to understand how a moral action is elicited and fostered such that volunteer tourism is an appropriate choice. Cottingham (2000) suggests that if we possess the character and personality to lead an ethical life, we may not be able to choose when to activate these traits in our character, and thus continuous cultivation of our ethical and emotional sensibilities is important in building up the virtue ethics to achieve an ethical life. When active strategies are available, these ethical traits are activated and individuals respond in the intended form of the strategies.

Habitus provides a perspective to explore the logic of practice through dispositions inculcated in the individuals' life trajectories. Bourdieu (1990b) defines habitus as:

> the conditionings associated with a particular class of conditions of existence produce habitus, systems of durable, transposable dispositions, structured structures predisposed to function as structuring structures, that is as principles which generate and organize practices and representations that can be objectively adapted to their outcomes without presupposing a conscious aiming at ends or an express mastery of the operations necessary in order to attain them.
>
> (p. 53)

It can be conceived as a scheme of perceptions and thoughts of action, through the process of socialisation, which provide an objective basis for the regularity of mode of behaviour. As a result, individuals act in a certain way in certain circumstances, in the way that the underlying structures are reproduced. The production of social practice is informed by the habitus internalised from the structural code of culture (Nash, 1999).

Habitus can also be conceived as what forms the basis of character of an individual, 'our own self' (Burkitt, 2002, p. 226). It is central to constituting humans into selves, and these dispositions are instilled from earlier years of education to produce a socialised, structured body (Burkitt, 2002; Reay, 2004). According to Bourdieu, habitus does not only constitute the self in the social world, but it also reflects how the immanent structures of the social world are incorporated in the body in certain ways that 'structures the perception of that world as well as action in that world' (Bourdieu, 1998, p. 81; as cited in Reay, 2004, p. 432). The practices produced by the habitus are unpredictable because habitus is internalised as a second nature and embodied as active presence of past experiences; it is also 'a spontaneity without consciousness or will' (Bourdieu, 1990b, p. 56), characterised by the logic 'of vagueness, of the more-or-less, which defines one's ordinary relation to the world' (Bourdieu, 1990a, p. 78). It offers a useful lens to understand the generative nature of habitus as a mediating construct to combine one's previous knowledge and experiences to produce particular responses to certain situations.

Besides producing practice, habitus also enables an individual to reproduce structures, generating a wider repertoire of possibility of actions and practices. It bears transformative potential that

> leads us to 'reproduce' the social conditions of our own production, but in a relatively unpredictable way, in which a way that one cannot move simply and mechanically from knowledge of the conditions of production to knowledge of the products.
>
> (Bourdieu, 1993, p. 87; as cited in Reay, 2004, p. 433)

So, habitus is responsive and permeable in the sense that circumstances exists not only to be acted upon but also to be internalised, adding another layer to the habitus acquired through earlier socialisation. Implicit in the concept is that habitus operates at an unconscious level until the individual encounters moments of self-questioning, at which point habitus operates at a conscious level in such a way that new facets of the self are developed (Reay, 2004).

As Bourdieu (1996) states, family is the primary space of accumulation and transmission of different forms of capital, including habitus (as cited in Nash, 1999). The habitus acquired in the family forms the basis of structuring experiences in other social institutions such as school; through socialisation in school, the habitus transformed by actions in the school in turn is reflected in subsequent experiences, 'from restructuring to restructuring' (Bourdieu, 1972; as cited in Reay, 2004, p. 434). As a result of socialisation in a different space, such as from family to school,

> members of a social group come to acquire a set of dispositions which reflect the central structural elements (such as political instability and kindship rules) of their society, and therefore behave in ways which necessarily reproduce those structural elements, although in a modified form.
>
> (Nash, 1999, p. 185)

This provides the basis for how patterns of thought emerge that challenge existing modes of thought or how individuals mediate the habitus which they have acquired.

Using Bourdieu's habitus is useful in understanding how participation in volunteer tourism and practices during the volunteering programme are informed by dispositions as the frames of reference which are internalised during the process of socialisation, and how the habitus serves as the basis upon which active strategies as well as value system of a particular culture play out. Bourdieu's emphasis on unconsciousness however directs our attention away from how the active strategies of 'deepening of personal responsibility' act upon the pattern of thought and elicit responsible action that an individual necessarily follows to achieve certain ends. The power of socialisation to impose social forms not only on thought but also on the physical behaviour of the body cannot be neglected. Concepts on conscious self-formation help to reconcile this limitation of habitus, contributing to the theoretical framework that explains the relationship between production of the moral self and choice made by the moral subject produced.

Quality and value accumulation to the moral self

Moving further from conscious self-making process and unconscious inculcation of dispositions that inform moral actions and practices, it is important to understand what quality or value the individual attains from or exhibits through moral actions and practices. *Suzhi*[8] (素質) is a particular and local notion of quality in the Great China Region and other Chinese-speaking countries in Asia. While habitus is unconsciously acquired, *suzhi* is the conscious accumulated value into the body which is embodied as certain level of quality (high or low) of an individual. It refers to the 'innate and nurtured physical, psychological, intellectual, moral, and ideological qualities of human bodies and their conduct' (Jacka, 2009, p. 524). *Suzhi*, as a form of quality, 'is not something that naturally inheres in the body but is rather something that must be built into the body' (Anagnost, 2004, p. 193). In other words, it is not a surplus value extracted from but something accumulated into the body. An individual is recognised as having value if value has been added to the body, a supplement to the 'bare life', through educational attainment or other strategies for the purpose of capital accumulation (Anagnost, 2004).

A *suzhi* discourse is found central particularly to contemporary China's governance and society in the understanding of responsibilities, obligations and rights that contribute to the cultivation of citizens. Accumulating *suzhi* to cultivate a moral self or to re-invent oneself into an individual with more values is an ordinary ethics. It has been discussed in relation to neoliberalism (see Anagnost, 2004; Yan, 2003). Kipnis (2007) posits that '*suzhi* discourse reifies rather than elides forms of hierarchical difference; it offers a way of speaking explicitly about class without using the word "class"' (p. 390). This 'blame the victim' but 'naturalise the victor' discourse is hierarchical and authoritarian rather than liberal or neoliberal. Nonetheless, the authoritarian means may result in governable subjects with deposited mode of practice. For instance, the change from compulsory and paid blood donation to steady increase in voluntary unpaid donation in China has demonstrated the shift from obligation to a norm, a 'wanted to' practice (Jun, 2011). This has helped to add value to the people at the same time the capital accumulation process has been internalised into the habitus, albeit the means was authoritarian at first.

In the Chinese society, self-cultivation is the process of becoming a moral person (Yan, 2013). The quality or character of an individual in Chinese culture is signified as *ren* (仁; translated as benevolence), which could be accumulated by taking up more roles and obligations. Having more *ren* makes the individual a moral exemplar – a role model of moral excellence in Confucianism who helps others in self-making and self-betterment (Humphreys, 2017). Exemplar morality is defined as 'a form of morality and social governance that seeks social order through leadership by example and emulation of role models' (Cody, 2018, p. 73). Exemplarity is one of the most effective strategies for self-cultivation in the principles in the Analects, and is still a visible characteristic of the Chinese moral landscape (Cody, 2018; Olberding, 2008).

The notion of *suzhi* is loosely or sometimes ambivalently in place in Hong Kong and Taiwan (research focus of this book) due to historical development that governance is not framed with such form of class distinction authoritarian approach commented by Kipnis (2007), but a higher degree of freedom and autonomy, especially experienced by the younger generation. The traditional Confucian values are still existing and important as a dominant system of social and ethical philosophy that governs everyday life. The essence of *suzhi* discourse as the active accumulation and enhancement of quality of an individual adds to the theoretical framework of this book in understanding the development and character of the moral self.

Moral self and the others

Having discussed about the cultivation of the 'self', or the 'moral self' with a focus on the 'self', it is essential to further unpack the 'self' in a social relation. 'Community' is a commonly used concept of social relations, helpful in discerning the meaning, constitution and development of the self in the process of socialisation. It can be conceptualised with two moral dimensions, first 'community is good in itself' and second 'it speaks with moral authority' (Smith, 1999, p. 20). Sandel (1982) explains that the sense of community does not only tell members 'what they have' but also 'what they are' (p. 150; as cited in Smith, 1999, p. 21), with the identity constituted through being in the community.

Etzioni (1995) defines community with a moral element, stating that 'communities are social webs of people who know one another as persons and have a moral voice' that helps to 'fill the moral vacuum' of contemporary society (p. ix). Under neoliberal individualism or market-driven competitiveness, 'moral fabric of communities was disintegrating' (Bowring, 1997, p. 96), resulting in a loss of trust, intimacy, mutuality and relationships in society, and then culminating alienation and social crises (Arai & Pedlar, 2003).

The communitarians call to 'restore civic virtues, for people to live up to their responsibilities and not merely focus on their entitlements... Communities draw on interpersonal bonds to encourage members to abide by shared values' (Etzioni, 1995, p. ix), bringing up the discussion of relational identity, 'we' versus 'I', and implying that the common good of the community comes before individual rights. Here, what Etzioni highlights is not on 'doing good', but on the 'common good' as a result of group endeavour with meaning to them (Arai & Pedlar, 2003). Communitarians regard humans as 'essentially social beings' and reaffirm the mutually supportive aspects of human life (Arai & Pedlar, 2003, p. 188; Frazer & Lacey, 1994). This makes the context in which collective values of reciprocity, solidarity and community become more apparent (Arai & Pedlar, 2003), and thus an ethic of care and responsibility follows naturally as people are connected (Haste, 1996).

Commonly, community is perceived as where people live and thus find meaningful interactions and social relations (Bradshaw, 2008). Globalisation makes it possible for social relations to be uncoupled from place as people are not required to live in proximity to engage in meaningful relations (Sasson, 2001; as cited in Bradshaw, 2008). The traditional place-based community is being eroded, and thus 'place' in question is getting less locally bounded, a more outward-looking 'global sense of place' (Massey, 1997) is displacing the intimate togetherness of local community (Smith, 1999). In response, Bradshaw (2008) argues for a concept of post-place community as alternative perspective, which sees 'community in terms of the networks of people tied together by solidarity, a shared identity and set of norms, that does not necessarily reside in a place' (p. 5). The shared identity means the ability to identify membership, to become a member, and to be recognised by the members of the community, which altogether is associated with belonging (Bradshaw, 2008).

Community constitutes a web of relationships which forms relationality which 'extends beyond simply relating to another; it also includes a state of humanness that co-exists with the experience of being in relationship' (Neville, 2008, p. 251). Community, deep in human aspirations, is an expression of the desire to experience a sense of belonging (Mulligan, Steele, Rickards, & Fünfgeld, 2016). Compared to the typical understanding of belonging as an intangible feeling, a sense of belonging, Bennett (2012) argues that belonging should be understood as practice, a way of being and acting in the world. To have 'a sense of belonging' is passive, intangible and ineffable while 'to belong' is to 'be a part of something, be accepted by others' which is active, tangible and recognised by others (Bennett, 2012, p. 31). These are cultural and embodied, about the practices and the ways of doing. Doing belonging can then be seen as a way of self-making, a process of bettering their relationship with others in society. It engages people in relations by taking up moral responsibility and reciprocity. This enters the state of 'being in a correct relation' (Miller, 2003, p. 218), which requires living in an ethical relationship with the self, others and the environment. This moral dimension of belonging entails reciprocated care that individuals do not exist independently of the society they live in

(Bennett, 2015) or not simply being related to others. In such, the self and society are 'mutually constitutive' (May, 2011; as cited in Bennett, 2012).

This mutuality puts forward Buber's (1958) concept of I–Thou relation to understand the mode of existence of human beings. According to Buber (1958), 'I' emerges from the dynamic relationship with others. The I–Thou relation is different from I–It relation in the sense that in the latter, others exist as objects fixed in space and time to be experienced while the former perceive others as subjects to encounter and experience with (Buber, 1958; Crossley, 1996). In other words, I–Thou relations are dialogical while I–It relations are monological; only through encounters in I–Thou relations can the self be developed into the whole being (Buber, 1958). The mutuality in I–Thou relations allows dialogic communities to be formed and evolved through encounters; it also allows the claiming of membership of a community to be accepted. On the other hand, I–It relations predominate the modern rationalised societies in which people exist as individual entities (Crossley, 1996). It is suggested that many people involved in post-place communities are better connected than those in traditional place-based communities, which would mean that the 'local vacuum' is less about a lack of community than a lack of interaction (Bradshaw, 2008).

Care is another key feature in the notion of community. Etzioni (1995) associates community with 'a place in which people know and care for one another' (p. 31). Care can be categorised into 'caring about' and 'caring for' (Silk, 2000); the former encompasses emotional engagement in the sense of emotional concern about others, and the latter involves active step of caring practice which also implies effective translation from ethical intention to actual behaviour (Barnett & Land, 2007). The 'landscape of care' helps to understand the geographical scales of caring relationships (Milligan & Wiles, 2010) and prompts us to explore different 'we' relations formed in various temporal spaces, particularly interpersonal bonds in communities, where caring relationships reflect interdependence (e.g. Arai & Pedlar, 2003; Held, 2006; Tronto, 1993). The ethics of care in communitarianism emphasise mutuality and a relational rather than autonomous identity (Smith, 1999). The spatial scale or spatiality of caring relationships is important in the sense that

> moral capacity is fostered, cultivated, and exercised within the social environment of the small-scale setting – or it is not acquired at all … what is fostered here, in the setting of proximity, is the capacity for developing empathy with others.
>
> (Vetlesen, 1993, p. 382; as cited in Smith, 1999)

The volunteer tourism literature tends to focus on caring for the distant others; there is scant research explicitly discussing the ethics of care in self-forming in a community formed in volunteer tourism. This tourism space provides a social context to actualise the meaning of moral obligation. In social contexts in which people feel engaged with and connected to other people, they are experiencing responsibility and caring/being cared which then become explicit and normative (Haste, 1996). I conceptualise this social context of moral development as a post-place community of volunteer tourism with the concept of liminality and the communitas produced in the collective space (Kwong, 2021). Liminality is a state in which wonder and realisation take place, a phase in our social life in which the travel between the structure and anti-structure produces moments of understanding who we are (Turner, 1969). Turner (1969) has proposed three stages of change – separation from the old, betwixt and between, and aggregation and return with a new status. All manner of possibilities could occur in the liminal, and people going through the stages could come out different (Turner, 2012). In this liminal space, what has been bounded by social structure in our everyday life is liberated, everyday practices and moral codes are disappearing or getting looser, with such characteristics as 'equality, undifferentiated humanness, androgyny, and humility' (Turner, 2012, p. 169). This liberation produces collective joy and unforgettable memories, a sense of comradeship and communitas – the state of being betwixt and between and people being treated equally (Turner, 2012; Turner, 1969). This sense of 'unaccommodated human being' (Turner, 2012) can be linked up with the sense of community, and the concept of 'I–Thou' relations by Buber (1958). Travelling comes as alternative rhythms to everyday life, resulting in different types of 'we' relations at different temporal spaces. This sense of community or communitas can be produced when people engage in collective tasks with dedication and commitment, resulting in enjoyment and at the same time a loss of ego (Turner, 1969). The collective joy and memories are believed to strengthen in time when people have to be apart, as Turner (2012) reiterates that communitas is not ephemeral, it folds over and back as it is 'not on our time scale' (p. 195). Kontogeorgopoulos (2017) explored the existential authenticity in motivations and their perceived benefits of international volunteer tourists in Thailand and explained the inter-personal existential authenticity using the concept of touristic communitas. Bonding and friendship among volunteers are developed when everyday norms, expectations and responsibilities are removed, interaction with strangers becomes easier, and 'inter-personal connections that are unlikely outside the liminal setting of volunteer tourism' are nurtured (Kontogeorgopoulos, 2017, p. 8). The liminal space provides a pleasant setting for finding oneself and understanding others when the stereotypes are suspended. This may then contribute to having a new experience of the self which results in different forms of the self. This then ties up the analytical framework by examining how the moral self is cultivated and experienced in relation to others, particularly through the communitas produced during the experience.

Researching non-Western volunteer tourists: A post-place ethnographic study of volunteers from Hong Kong and Taiwan

This book is based on a six-month post-place ethnography in 2016–2017 in Hong Kong, Taiwan and Cambodia. Ethnography is a methodological choice which 'privileges an engaged, contextually rich and nuanced type of qualitative social research, in which fine grained daily interactions constitute the lifeblood of the data produced' (Falzon, 2009, p. 1). It allows the researcher to capture the social meanings of practices and ordinary activities of the researched in a particular setting of 'field', through participating directly in that setting, in order to collect data in a systematic manner but 'without meaning being imposed on them externally' (Brewer, 2000, p. 11). The value of ethnography has been recognised within 'everyday' studies to understand how people live their everyday lives out (Crang & Cook, 2007). With the aim of moving the discussion beyond the non-Western context, this study focused on Chinese volunteers originally from Hong Kong and Taiwan in East Asia using participant observation and interviews. 'Chinese' here refers to the Chinese culture and its traditions and customs with over 5000 years of development, and thus people with this cultural background rather than the Chinese nationals, *'hua ren 華人'* instead of *'zhong guo ren 中國人'*. Besides aligning with the objective of this book to explore the cultural specificity of the rationalities and practices, this clarification also seeks to avoid any political sensitivity due to recent political controversies between China, Hong Kong and Taiwan. For specific Chinese phrases, I use Mandarin pinyin and Traditional Chinese characters in addition to English translation to make it understandable to all readers.

Adopting the multi-sited approach – by involving volunteers from two origins – is not for comparison purpose. A multi-sited ethnography 'involves a spatially dispersed field through which the ethnographer moves – actually, via sojourns in two or more places, or conceptually, by means of techniques of juxtaposition of data' (Falzon, 2009, p. 2). It offers a more holistic perspective and helps to address some practical restrictions in recruiting informants. The group size of voluntary service trips from these two origins is usually small, around 10–20 participants, and the duration of the programmes is short, around 5–21 days. Involving two groups of volunteer tourists enhances the breadth and depth of the data, as there are differences in the development of volunteerism in Hong Kong and Taiwan despite their cultural similarities. The recruitment of informants was restricted by the availability of service trips (usually more available during summer and winter breaks), the schedule of the programmes that allowed me to conduct ethnographic study, and limited time and resources of this study.

With the rise and increasing popularity of the notions of global citizenship, service learning, and experiential learning in schools and universities in Hong Kong, a culture of giving and social responsibility is widely promoted. Strategic philanthropic activity is thriving among young people, as they tend to move away from the traditional way of sending a cheque to become more proactive and engaged in philanthropy through volunteer tourism (Cheung, 2016). Many service trips and overseas community projects are organised by student organisations in collaboration with NGOs.

Volunteerism in Taiwan could be traced back to the 1970s when its economy developed rapidly as a way of 'giving back' to the international aid and assistance received in the 1960s. The government has set up and encouraged the establishment of NGOs to promote public service from the 1970s onwards, including the International Cooperation and Development Fund (TaiwanICDF) which aims at socio-economic development, enhancing human resources, and promoting economic relations in a range of developing partner countries (TaiwanICDF, 2004). Following this, various NGOs have initiated overseas community service projects and service learning has become a compulsory element of the curriculum. The international volunteer tourism phenomenon in Taiwan has aroused public attention and attracted open debate of its values, influences, and impacts (Tu, 2011).

Before approaching potential informants for my research, two sending organisations – The Bucket Wish (TBW) in Hong Kong and ELIV International Service (ELIV) in Taiwan – were contacted for collaboration in recruiting research informants. TBW is a student-led not-for-profit organisation founded by three university students in 2013 focusing on community-based sustainable development primarily in Cambodia. The organisation is sustained on project fees paid by participants, and additional funding from the Centre of Development and Resources for Students in HKU and a leading youth development NGO in Hong Kong, Youth Global Network of Breakthrough Hong Kong. The project coordinators of TBW are students who volunteer for both planning and organising the project and the community service work in Cambodia. Each year in January, TBW organises two parallel service project teams in Cambodia with local partners. Before this study, I participated in one of its service trips as a volunteer and project coordinator.

Established in 2010, ELIV is one of the first social enterprises in Taiwan. They offer a wide range of voluntary projects throughout the year aiming for empowering learning through innovative volunteerism, improvements through understanding social issues and needs as well as community development in different Asian countries including Cambodia, Nepal, Myanmar, China and Taiwan. Administrative costs and community development expenses such as construction materials are covered by programme fees. ELIV International Service Association was formed in 2012 as a partnering NGO only to receive donation to support the projects. Project Cambodia was initiated in 2012 and managed by Cambodia-based Taiwanese staff, and later led by a Cambodian Project Manager and Assistant Manager.

I signed up as a volunteer in one project organised by these two organisations to become a 'participant as observer' who is an ethnographer with a high level of participation in the researched community (Gold, 1958). As Paul (1953) has pointed out, 'participation implies emotional involvement; observation requires detachment. It is a strain to try to sympathise with others and at the same time strive for scientific objectivity' (as cited in Musante & DeWalt, 2010, p. 28). This ethnographic role engaged me into the same time and space with the post-place communities of volunteer tourists for participant observation so that I had a shared identity with the volunteers and become an insider of the group. Therefore, it was not just about 'watching'; I did the same as other volunteers required for the project and had genuine interactions with the team members as part of the post-place community, at the same time jotted down observation notes as a researcher. Spending time with the volunteers during pre-trip preparation, discussion and reflection, voluntary work, and informal interactions such as eating together and commuting all helped to develop understanding through involving in the spontaneity of everyday interactions (Griffiths, 2015). This facilitated me to ask meaningful and relevant questions during informal communications and personal interviews. Throughout the study, I conducted pre-trip and post-trip interviews with 34 volunteers aged 16–61, of which 21 were from Hong Kong and 13 were from Taiwan. The interviews were conducted in person in Hong Kong and Taiwan, except a few pre-trip interviews with the Taiwanese informants that were conducted online (see Kwong, 2021 for more details about the recruitment of research participants). All voices are anonymised with pseudonyms.

Two volunteering projects

I joined the 'Project A-Mangrove Planting x Community Mapping' in Cambodia (hereafter TBW-Team A) organised together by TBW and a local NGO Cambodia Youth Action (CYA) during 4–15 January 2017. This programme was largely based in the Trapeang Sangke Community, Kampot Province for 10 days to volunteer for the mangrove restoration project initiated by the local fishers. The villagers collaborated with CYA to recruit international volunteers to restore the lost mangrove forest due to land clearance for private development. Volunteers were scheduled to collect seeds from the mangrove seeds in the sea, arranged them in the seedling nursery site, and planted seedlings prepared by previous volunteer groups into the sea. The itinerary also included constructing part of a footbridge along the river, community mapping to understand everyday life, cultures, and conditions of the local community (partly to provide updates to CYA for planning), an inspiration day to engage kids in stimulating creativity through fun games, a cultural night and a cooking challenge. Discussion, debriefing and reflection sessions were facilitated by CYA's camp leader or TBW's project coordinators. We had visits to the Tuol Sleng Genocide Museum (also known as S-21) and Choeung Ek Killing Fields on the first two days and a free day on the last day of the service trip in Phnom Penh. I also visited the parallel project team TBW-Team B in Tanoun Village in Takeo Province on 10–11 January with the TWB Chairperson. This allowed me to participate in the volunteering programme and interactions among the project team as well as with the locals. Besides the time in Cambodia, participants were required to attend four pre-trip workshops for topical and contextual discussions and preparing work materials. Four post-trip workshops were also organised to reflect on issues such as migrant workers and ethnic minority discussed in Cambodia but in the Hong Kong context.

I then joined Project Cambodia organised by ELIV, which lasted for eight days in Siem Reap during 4–11 February 2017. The Project emphasises sustainable community development by engaging local community and training local staff, and understanding of local issues through visits and discussions. The project has taken place in Svay Chek Village since 2013, with fixed activities for each project team including building houses and toilets, promoting hygiene education, designing short activities for the primary school students, and an outing to the Angkor UNESCO Heritage Site (hereafter the Angkor) with some of the students who had never been there. For the project team I participated in, we also had home visits in the village and other visits such as the Kamonohashi Project which is a social enterprise founded to help victims of domestic violence, charity restaurants and market, and Wat Thmei Killing Field. Discussion, reflection and sharing sessions took place almost every evening moderated by two project leaders, the Project Director and/or the Project Manager. I attended a briefing and preparation session organised on 24 December 2016 in Taipei City, after which the project team discussed online about the school activities.

Analysis

While I have reflected on my identities and roles as well as the relational ethics embedded that framed my practices as a fieldworker in this ethnographic process in another occasion (see Kwong, 2021), I want to reflect here on the challenges from presenting the data. All the interviews and interactions were conducted in Cantonese and Mandarin, and translated into English for coding and analysis. As this study looks at the cultural specificity, I was very cautious in interpreting the meaning of the same phrase in different contexts and translating it into English as accurately as possible. During transcription and translation, I used the same pinyin expressions and included the meaning and Chinese characters in brackets for some phrases which have been commonly used in the literature, such as *suzhi*. When I found phrases that I could not find a direct translation, I included the Chinese phrases in brackets next to the translation. I also kept a list of literal translation so as to standardise the translation across transcripts. For some phrases which could have several meanings or translation, I translated them according to the context, and put down all the translations of the same phrase on the list. The difficulty in translating the expressions from Chinese to English precisely explains the uniqueness of values and ethics in non-Western contexts.

The process of analysis was driven by the data rather than imposing an analytical frame on the data from the beginning. As Dixon-Woods, Fitzpatrick, and Roberts (2001) state, 'qualitative data take the form of narrative, with themes and concepts as the analytical device' (p. 126). This helped to lay out a flow of quotes built upon the linkages between responses and stories. This approach helps to unveil the specific rather than the abstract, to reveal the languages and their nuances in the unfolded experience in a temporal way and to allow the personal dynamics to reveal themselves in the practices and relationships constructed (Jones, 2004). Therefore, in the following chapters, I tend to weave the analysis and discussion with quite a lot of direct quotes from the informants, as their voices speak more directly and authentically to the contexts and experiences. By 'voice' here I refer to the character central to the story reflecting what the story was and how the story was told. This aims to move the analysis away from situating the researcher as the storyteller with an invisible voice along the writing, to forging a conversation among real voices of the informants while the researcher as the participant was also part of the conversation so that the voices were represented with justice (Nunan & Choi, 2011). For instance, Camellia's story was so intense and remarkable that several informants shared their feelings and reflections when hearing about the story in our post-trip interviews (*Chapter 3*). Hence, I used Camellia's voice to start the discussion which was attempted to understand how this particular social setting helps to reveal the authentic self in self-making. Other voices were brought in to bring out broader issues such as their traditional values have shaped the relations of the self with oneself and others. This created a dialogue among the informants to piece together the narrative or various narratives to reveal and reflect on the issues and debates concerned, while my role was to interpret their opinions and feelings with reference to the analytical framework and my understanding of the culture. This also offers authentic materials for further or different interpretations and opening up more critical discussions.

Synopsis

After situating volunteer tourism and the moral self in the analytical framework presented in this chapter, *Chapter 2* is premised on the question of how volunteer tourism has become an appropriate choice. By conceiving volunteer tourism as a social trend, it unravels the moral landscape in Hong Kong and Taiwan to understand the interplay of different values in rationalising this choice of activity for a class of moral subjects. It also attends to the cultural specificity of the rationalities.

Chapter 3 explores how and what kinds of relationships were produced through the lens of liminality. I propose that three forms of moral communities are constructed or sustained by the volunteers through their practices of doing belonging. This chapter then discusses how ethics of care, responsibility and obligation are enacted in a liminal tourism space by looking at different 'we' relations. It aims to understand how the moral self is cultivated and experienced in relation to others.

Chapter 4 attempts to unpack how transformation can be facilitated and what kinds of transformation are possible before concluding that transformation is a definite outcome of travelling as a volunteer. It focuses on how the moral self is developed from the activities and practices during the volunteering programme. I focus on the spatial and temporal dimensions of the transformational process to explore what changes are possible beyond immediate in-situ impacts.

The discussion on the potentiality of transformation through volunteer tourism continues in *Chapter 5* by looking at how the sending organisations play a crucial role in reshaping the orientations of the moral self. Pre- and post-trip workshops, debriefings during the trip, and injecting a theme to activities could also be seen as tools for disorienting dilemma, introspecting, and re-orienting for changes. Meanwhile, there is arguably a risk that their acts of travelling ethically tend to be rationalised instead of being encountered naturally.

In *Chapter 6*, I draw together the main lines of argument across the book to re-conceptualise volunteer tourism based on the 'what' and 'why' of motivations and outcomes. It also calls for more contributions with non-Western and interdisciplinary perspectives to the volunteer tourism literature.

Notes

1 Examples include: Benson and Seibert (2009); Broad and Jenkins (2008); Bruyere and Rappe (2007); Chen and Chen (2011); Grimm and Needham (2012); Lo and Lee (2011); Sin (2009); Söderman and Snead (2008); Tomazos and Butler (2010); Weaver (2015).
2 Examples include: Coghlan (2015); Conran (2011); Frazer and Waitt (2016); Guttentag (2009); Holmes and Smith (2010); Lepp (2008); McGehee and Andereck (2009); McGehee and Santos (2005); Mostafanezhad (2016); Mostafanezhad, Azizi, and Johansen (2016); Zahra and McGehee (2013).
3 Examples include: Butcher (2003); Griffiths and Brown (2016); McGehee and Andereck (2008); Palacios (2010); Sin (2009, 2010); Theerapappisit (2009).
4 Examples include: Caton and Santos (2009); Guttentag (2009); Lyons et al. (2012); Palacios (2010).
5 Examples include: Baillie Smith and Laurie (2011); Baillie Smith, Laurie, and Griffiths (2018); Laurie and Baillie Smith (2018); Mostafanezhad (2012); Noxolo, Raghuram, and Madge (2012); Simpson (2004).
6 Examples include: Baillie Smith and Laurie (2011); Burrai, Mostafanezhad, and Hannam (2017); Griffiths (2015); Mostafanezhad (2014); Vrasti (2013).
7 Examples include: Ateljevic and Tomljenovic (2016); Coghlan and Gooch (2011); Coghlan and Weiler (2018); Müller, Scheffer, and Closs (2020); Pritchard et al. (2011); Tomazos and Murdy (2023).
8 It is termed *zhisu* (質素) in Hong Kong (Cantonese speaking). Similar terms for quality could include *xiuyang* (修養, cultivation) and *renpin* (人品, character or moral quality).

References

Allahyari, R. A. (2000). *Visions of charity: Volunteer workers and moral community*. London: University of California Press.
Anagnost, A. (2004). The corporeal politics of quality (Suzhi). *Public Culture*, *16*(2), 189 208.
Arai, S., & Pedlar, A. (2003). Moving beyond individualism in leisure theory: A critical analysis of concepts of community and social engagement. *Leisure Studies*, *22*(3), 185–202. doi:10.1080/026143603200075489
Ateljevic, I. (2009). Transmodernity: Remaking our (tourism) world. In J. Tribe (Ed.), *Philosophical issues in tourism* (pp. 278–300). Bristol: Channel View Publications.

Ateljevic, I., & Tomljenovic, R. (2016). Triple T: Tourism, transmodernity and transformative learning. *Turističko Poslovanje, 2016*(17), 25–31.
Baillie Smith, M., & Laurie, N. (2011). International volunteering and development: Global citizenship and neoliberal professionalisation today. *Transactions of the Institute of British Geographers, 36*(4), 545–559. doi:10.1111/j.1475-5661.2011.00436.x
Baillie Smith, M., Laurie, N., & Griffiths, M. (2018). South–South volunteering and development. *The Geographical Journal, 184*(2), 158–168. doi:10.1111/geoj.12243
Barnett, C., Cloke, P., Clarke, N., & Malpass, A. (2005). Consuming ethics: Articulating the subjects and spaces of ethical consumption. *Antipode, 37*(1), 23–45.
Barnett, C., Cloke, P., Clarke, N., & Malpass, A. (2010a). The ethical problematization of 'the consumer'. In C. Barnett, P. Cloke, N. Clarke, & A. Malpass (Eds.), *Globalizing responsibility: The political rationalities of ethical consumption* (pp. 27–60). Chichester, West Sussex: Wiley-Blackwell.
Barnett, C., Cloke, P., Clarke, N., & Malpass, A. (Eds.) (2010b). *Globalizing responsibility: The political rationalities of ethical consumption*. Chichester, West Sussex: Wiley-Blackwell.
Barnett, C., & Land, D. (2007). Geographies of generosity: Beyond the 'moral turn'. *Geoforum, 38*(6), 1065–1075. doi:10.1016/j.geoforum.2007.02.006
Beck, U. (1992). *Risk society: Towards a new modernity* (Vol. 17). London: Sage.
Bennett, J. (2012). *Doing belonging: A sociological study of belonging in place as the outcome of social practices*. PhD thesis. University of Manchester, UK.
Bennett, J. (2015). 'Snowed in!': Offbeat rhythms and belonging as everyday practice. *Sociology, 49*(5), 955–969.
Benson, A., & Seibert, N. (2009). Volunteer tourism: Motivations of German participants in South Africa. *Annals of Leisure Research, 12*(3–4), 295–314. doi:10.1080/11745398.2009.9686826
Boluk, K. A. (2011). Fair Trade Tourism South Africa: Consumer virtue or moral selving? *Journal of Ecotourism, 10*(3), 235–249. doi:10.1080/14724049.2011.617451
Bourdieu, P. (1990a). *In other words: Essays towards a reflexive sociology*. Stanford, CA: Stanford University Press.
Bourdieu, P. (1990b). *The logic of practice*. Stanford, CA: Stanford University Press.
Bourdieu, P. (1993). *Sociology in question*. London: Sage.
Bowring, F. (1997). Communitarianism and morality: In search of the subject. *New Left Review* (222), 93.
Bradshaw, T. K. (2008). The post-place community: Contributions to the debate about the definition of community. *Community Development, 39*(1), 5–16.
Brewer, J. D. (2000). *Ethnography*. Buckingham: Open University Press.
Broad, S., & Jenkins, J. (2008). Gibbons in their midst? Conservation volunteers' motivations at the Gibbon Rehabilitation Project, Phuket, Thailand. In K. D. Lyon & S. Wearing (Eds.), *Journeys of discovery in volunteer tourism: International case study perspectives* (pp. 72–85). Cambridge, MA: CABI Publishing.
Bruyere, B., & Rappe, S. (2007). Identifying the motivations of environmental volunteers. *Journal of Environmental Planning and Management, 50*(4), 503–516. doi:10.1080/09640560701402034
Buber, M. (1958). *I and thou* (R. G. Smith, Trans.). Edinburgh: T&Clark Ltd.
Burchell, G. (1993). Liberal government and techniques of the self. *Economy and Society, 22*(3), 267–282. doi:10.1080/03085149300000018
Burkitt, I. (2002). Technologies of the self: Habitus and capacities. *Journal for the Theory of Social Behaviour, 32*(2), 219–237.
Burrai, E., Mostafanezhad, M., & Hannam, K. (2017). Moral assemblages of volunteer tourism development in Cusco, Peru. *Tourism Geographies, 19*(3), 362–377. doi:10.1080/14616688.2016.1236145
Butcher, J. (2003). *The moralisation of tourism: Sun, sand … and saving the world?* London: Routledge.

Butcher, J. (2014). Moralizing tourism: Personal qualities, political issues. In M. Mostafanezhad & K. Hannam (Eds.), *Moral encounters in tourism* (pp. 17–29). Surrey: Ashgate.

Butcher, J., & Smith, P. (2010). 'Making a Difference': Volunteer tourism and development. *Tourism Recreation Research*, *35*(1), 27–36. Retrieved from http://research.smuc.ac.uk/id/eprint/280

Caton, K. (2012). Taking the moral turn in tourism studies. *Annals of Tourism Research*, *39*(4), 1906–1928. doi:10.1016/j.annals.2012.05.021

Caton, K., & Santos, C. A. (2009). Images of the other: Selling study abroad in a postcolonial world. *Journal of Travel Research*, *48*(2), 191–204.

Centre for the Promotion of Imports from developing countries (CBI). (2020, 13 January). The European market potential for SAVE tourism. Retrieved from https://www.cbi.eu/market-information/tourism/save-tourism/market-potential

Chen, J. (2018). Understanding development impact in international development volunteering: A relational approach. *The Geographical Journal*, *184*(2), 138–147.

Chen, L.-J., & Chen, J. S. (2011). The motivations and expectations of international volunteer tourists: A case study of "Chinese Village Traditions". *Tourism Management*, *32*(2), 435–442. doi:10.1016/j.tourman.2010.01.009

Cheung, R. (2016, 11th August, 2016). Why Hong Kong students' volunteering may do more harm than good. *South China Morning Post*. Retrieved from http://www.scmp.com/lifestyle/travel-leisure/article/2001810/why-hong-kong-students-volunteering-may-do-more-harm-good

Cloke, P., Johnsen, S., & May, J. (2007). Ethical citizenship? Volunteers and the ethics of providing services for homeless people. *Geoforum*, *38*(6), 1089–1101. doi:10.1016/j.geoforum.2006.07.005

Cody, S. (2018). Borrowing from the rural to help the urban: Organic farming exemplars in postsocialist China. *The Asia Pacific Journal of Anthropology*, *19*(1), 72–89. doi:10.1080/14442213.2017.1394362

Coghlan, A. (2015). Prosocial behaviour in volunteer tourism. *Annals of Tourism Research*, *55*, 46–60. doi:10.1016/j.annals.2015.08.002

Coghlan, A., & Gooch, M. (2011). Applying a transformative learning framework to volunteer tourism. *Journal of Sustainable Tourism*, *19*(6), 713–728. doi:10.1080/09669582.2010.542246

Coghlan, A., & Weiler, B. (2018). Examining transformative processes in volunteer tourism. *Current Issues in Tourism*, *21*(5), 567–582. doi:10.1080/13683500.2015.1102209

Conran, M. (2011). They really love me!: Intimacy in volunteer tourism. *Annals of Tourism Research*, *38*(4), 1454–1473. doi:10.1016/j.annals.2011.03.014

Cottingham, J. (2000). Caring at a distance: (Im)partiality, moral motivation and the ethics of representation - partiality, distance and moral obligation. *Ethics, Place & Environment*, *3*(3), 309–313. doi:10.1080/713665894

Couch, J., & Georgeou, N. (2017). Do immersion tours have long-term transformative impacts on students?: A study of Australian university students in a Tibetan host community in India. *Journal of Applied Youth Studies*, *2*(2), 18.

Crang, M., & Cook, I. (2007). *Doing ethnographies*. London: SAGE Publications Ltd.

Crang, M., & Hughes, A. (2015). Globalizing ethical consumption. *Geoforum*, *67*, 131–134. doi:10.1016/j.geoforum.2015.10.005

Crossley, N. (1996). *Intersubjectivity: The fabric of social becoming*. London: SAGE Publications.

Daya, S. (2016). Ordinary ethics and craft consumption: A southern perspective. *Geoforum*, *74*, 128–135. doi:10.1016/j.geoforum.2016.06.005

Dixon-Woods, M., Fitzpatrick, R., & Roberts, K. (2001). Including qualitative research in systematic reviews: Opportunities and problems. *Journal of Evaluation in Clinical Practice*, *7*(2), 125–133. doi:10.1046/j.1365-2753.2001.00257.x

Etzioni, A. (1995). *The spirit of community: Rights, responsibilities and the communitarian agenda*. London: Fontana.
Falzon, M.-A. (2009). *Multi-sited ethnography: Theory, praxis and locality in contemporary research*. Farnham: Ashgate.
Feighery, W. G. (2011). Consulting ethics. *Annals of Tourism Research*, *38*(3), 1031–1050. doi:10.1016/j.annals.2011.01.016
Fornet-Betancourt, R., Becker, H., Gomez-Müller, A., & Gauthier, J. (1987). The ethic of care for the self as a practice of freedom: An interview with Michel Foucault on January 20, 1984. *Philosophy & Social Criticism*, *12*(2–3), 112–131.
Foucault, M. (1985). Morality and practice of the self (R. Hurley, Trans.). In *The history of sexuality Vol. 2: The use of pleasure* (pp. 2533). New York: Pantheon Books.
Foucault, M. (1988). Technologies of the self. In H. G. Luther, H. Martin, & P. H. Hutton (Eds.), *Technologies of the self: A seminar with Michel Foucault* (pp. 16–49). Massachusetts: University of Massachusetts Press.
Frazer, E., & Lacey, N. (1994). The politics of community: A feminist critique of the liberal-communitarian debate. *Cambridge Law Journal*, *53*, 619–619.
Frazer, R., & Waitt, G. (2016). Pain, politics and volunteering in tourism studies. *Annals of Tourism Research*, *57*, 176–189. doi:10.1016/j.annals.2016.01.001
Garthwaite, K. (2017). 'I feel I'm Giving Something Back to Society': Constructing the 'Active Citizen' and responsibilising foodbank use. *Social Policy and Society*, *16*(2), 283–292. doi:10.1017/S1474746416000543
Gharib, M. (2021, July 15). The pandemic changed the world of 'Voluntourism.' some folks like the new way better. Retrieved from https://www.npr.org/sections/goatsandsoda/2021/07/15/1009911082/the-pandemic-changed-the-world-of-voluntourism-some-folks-like-the-new-way-bette
Gold, R. L. (1958). Roles in sociological field observations. *Social Forces*, *36*(3), 217–223.
Griffiths, M. (2015). I've got goose bumps just talking about it!: Affective life on neoliberalized volunteering programmes. *Tourist Studies*, *15*(2), 205–221. doi:10.1177/1468797614563437
Griffiths, M., & Brown, E. J. (2016). Embodied experiences in international volunteering: Power-body relations and performative ontologies. *Social & Cultural Geography*, *18*(5), 665–682. doi:10.1080/14649365.2016.1210670
Grimm, K. E., & Needham, M. D. (2012). Moving beyond the "I" in motivation: Attributes and perceptions of conservation volunteer tourists. *Journal of Travel Research*, *51*(4), 488–501. doi:10.1177/0047287511418367
Grusky, S. (2000). International service learning: A critical guide from an impassioned advocate. *American Behavioral Scientist*, *43*(5), 858–867.
Guttentag, D. A. (2009). The possible negative impacts of volunteer tourism. *International Journal of Tourism Research*, *11*(6), 537–551. doi:10.1002/jtr.727
Hall, S. M. (2011). Exploring the 'ethical everyday': An ethnography of the ethics of family consumption. *Geoforum*, *42*(6), 627–637. doi:10.1016/j.geoforum.2011.06.009
Hammersley, L. A. (2014). Volunteer tourism: Building effective relationships of understanding. *Journal of Sustainable Tourism*, *22*(6), 855–873.
Haste, H. (1996). Communitarianism and the social construction of morality. *Journal of Moral Education*, *25*(1), 47–55.
Held, V. (2006). *The ethics of care: Personal, political, and global*. New York: Oxford University Press.
Henry, J. (2019). Directions for volunteer tourism and radical pedagogy. *Tourism Geographies*, *21*(4), 561–564. 10.1080/14616688.2019.1567577
Holmes, K., & Smith, K. (2010). The outcomes of tourism volunteering. In K. Holmes & K. Smith (Eds.), *Managing volunteers in tourism* (pp. 47–63). Oxford: Butterworth-Heinemann.

Humphreys, J. (2017, July 7). What does the rise of China mean for western values? *Irish Times*. https://www.irishtimes.com/culture/what-does-the-rise-of-china-mean-for-western-values-1.3276600

Jacka, T. (2009). Cultivating citizens: Suzhi (quality) discourse in the PRC. *Positions: Asia Critique*, *17*(3), 523–535. doi:10.1215/10679847-2009-013

Jones, K. (2004). Mission drift in qualitative research, or moving toward a systematic review of qualitative studies, moving back to a more systematic narrative review. *The Qualitative Report*, *9*(1), 94–111.

Jun, J. (2011). From commodity of death to gift of life. In A. Kleinman, Y. Yan, & J. Jun (Eds.), *Deep China: The moral life of the person* (pp. 78–105). Berkely: University of California Press.

Kelly, P. (2001). Youth at risk: Processes of individualisation and responsibilisation in the risk society. *Discourse: Studies in the Cultural Politics of Education*, *22*(1), 23–33. doi:10.1080/01596300120039731

Kipnis, A. (2007). Neoliberalism reified: Suzhi discourse and tropes of neoliberalism in the People's Republic of China. *Journal of the Royal Anthropological Institute*, *13*(2), 383–400. doi:10.1111/j.1467-9655.2007.00432.x

Kontogeorgopoulos, N. (2017). Finding oneself while discovering others: An existential perspective on volunteer tourism in Thailand. *Annals of Tourism Research*, *65*, 1–12. doi:10.1016/j.annals.2017.04.006

Kwong, Y. M. C. (2021). Travelling with the field: Post-place communities of volunteer touristsvolunteer tourists on the move. In M. O. Ajebon, Y. M. C. Kwong, & D. Astorga de Ita (Eds.), *Navigating the field: Postgraduate experiences in social research* (pp. 25–35). Cham: Springer International Publishing.

Lambek, M. (2010a). *Ordinary ethics: Anthropology, language, and action*. New York: Fordham University Press.

Lambek, M. (2010b). Toward an ethics of the act. In M. Lambek (Ed.), *Ordinary ethics: Anthropology, language, and action* (pp. 39–63). New York: Fordham University Press.

Laurie, N., & Baillie Smith, M. (2018). Unsettling geographies of volunteering and development. *Transactions of the Institute of British Geographers*, *43*(1), 95–109. doi:10.1111/tran.12205

Lee, S., & Yen, C.-L. (2015). Volunteer tourists' motivation change and intended participation. *Asia Pacific Journal of Tourism Research*, *20*(4), 359–377. doi:10.1080/10941665.2014.898672

Lepp, A. (2008). Discovering self and discovering others through the Taita Discovery Centre Volunteer Tourism Programme, Kenya. In K. D. Lyon & S. Wearing (Eds.), *Journeys of discovery in volunteer tourism: International case study perspectives* (pp. 86–100). Cambridge, MA: CABI Publishing.

Lo, A. S., & Lee, C. Y. S. (2011). Motivations and perceived value of volunteer tourists from Hong Kong. *Tourism Management*, *32*(2), 326–334. doi:10.1016/j.tourman.2010.03.002

Lyons, K., Hanley, J., Wearing, S., & Neil, J. (2012). Gap year volunteer tourism: Myths of global citizenship? *Annals of Tourism Research*, *39*(1), 361–378. doi:10.1016/j.annals.2011.04.016

Massey, D. (1997). A global sense of place. In T. J. B. D. Gregory (Ed.), *Reading human geography: The poetics and politics of inquiry* (pp. 315–323). London; New York: Arnold.

McGehee, N. G. (2014). Volunteer tourism: Evolution, issues and futures. *Journal of Sustainable Tourism*, *22*(6), 847–854. doi:10.1080/09669582.2014.907299

McGehee, N. G., & Andereck, K. (2008). 'Pettin'the critters': Exploring the complex relationship between volunteers and the voluntoured in McDowell County, West Virginia, USA, and Tijuana, Mexico. In K. D. W. Lyons & S. Wearing (Eds.), *Journeys of discovery in volunteer tourism: International case study perspectives* (pp. 12–24). Wallingford: CABI.

McGehee, N. G., & Andereck, K. (2009). Volunteer tourism and the "voluntoured": The case of Tijuana, Mexico. *Journal of Sustainable Tourism*, *17*(1), 39–51. doi:10.1080/09669580802159693

McGehee, N. G., & Santos, C. A. (2005). Social change, discourse and volunteer tourism. *Annals of Tourism Research*, *32*(3), 760–779. doi:10.1016/j.annals.2004.12.002

Mezirow, J. (1991). *Transformative dimensions of adult learning*. San Francisco, CA: Jossey-Bass.

Miller, L. (2003). Belonging to country – a philosophical anthropology. *Journal of Australian Studies*, *27*(76), 215–223.

Milligan, C., & Wiles, J. (2010). Landscapes of care. *Progress in Human Geography*, *34*(6), 736–754. doi:10.1177/0309132510364556

Mostafanezhad, M. (2012). The geography of compassion in volunteer tourism. *Tourism Geographies*, *15*(2), 318–337. doi:10.1080/14616688.2012.675579

Mostafanezhad, M. (2014). *Volunteer tourism: Popular humanitarianism in neoliberal times*. London: Routledge.

Mostafanezhad, M. (2016). Organic farm volunteer tourism as social movement participation: A Polanyian political economy analysis of World Wide Opportunities on Organic Farms (WWOOF) in Hawai'i. *Journal of Sustainable Tourism*, *24*(1), 114–131. doi:10.1080/09669582.2015.1049609

Mostafanezhad, M., Azizi, S., & Johansen, K. (2016). Valuing organic farm volunteer tourists in Hawai'i: Farm host perspectives. *Current Issues in Tourism*, *19*(6), 613–617. doi:10.1080/13683500.2014.961904

Mostafanezhad, M., & Hannam, K. (Eds.). (2014). *Moral encounters in tourism*. London: Routledge.

Müller, C. V., Scheffer, A. B. B., & Closs, L. Q. (2020). Volunteer tourism, transformative learning and its impacts on careers: The case of Brazilian volunteers. *International Journal of Tourism Research*, *22*(6), 726–738.

Mulligan, M., Steele, W., Rickards, L., & Fünfgeld, H. (2016). Keywords in planning: What do we mean by 'community resilience'? *International Planning Studies*, *21*(4), 348–361. doi:10.1080/13563475.2016.1155974

Musante, K., & DeWalt, B. R. (2010). *Participant observation: A guide for fieldworkers*. Lanham: AltaMira Press.

Nash, R. (1999). Bourdieu, 'Habitus', and Educational Research: Is it all worth the candle? *British Journal of Sociology of Education*, *20*(2), 175–187. doi:10.1080/01425699995399

Neville, M. G. (2008). When poor is rich: Transformative power of I-Thou relationships in a Brazilian favela. *Gestalt Review*, *12*(3), 248–266.

Noxolo, P., Raghuram, P., & Madge, C. (2012). Unsettling responsibility: Postcolonial interventions. *Transactions of the Institute of British Geographers*, *37*(3), 418–429. doi:10.1111/j.1475-5661.2011.00474.x

Nunan, D., & Choi, J. (2011). Shifting sands: The evolving story of "Voice" in qualitative research. In E. Hinkel (Ed.), *Handbook of research in second language teaching and learning* (pp. 222–236). Routledge.

Olberding, A. (2008). Dreaming of the Duke of Zhou: Exemplarism and the Analects. *Journal of Chinese Philosophy 35*(4), 625–639. doi:10.1111/j.1540-6253.2008.00508.x

Ong, F., King, B., Lockstone-Binney, L., & Junek, O. (2018). Going global, acting local: Volunteer tourists as prospective community builders. *Tourism Recreation Research*, *43*(2), 135–146.

Palacios, C. M. (2010). Volunteer tourism, development and education in a postcolonial world: Conceiving global connections beyond aid. *Journal of Sustainable Tourism*, *18*(7), 861–878.

Pan, T.-J. (2012). Motivations of volunteer overseas and what have we learned – The experience of Taiwanese students. *Tourism Management*, *33*(6), 1493–1501. doi:10.1016/j.tourman.2012.02.003

Pan, T.-J. (2017). Personal transformation through volunteer tourism: The evidence of Asian students. *Journal of Hospitality & Tourism Research*, *41*(5), 609–634.
Polus, R., Carr, N., & Walters, T. (2023). Challenging the eurocentrism in volunteering. *World Leisure Journal*, *65*(1), 101–118. doi:10.1080/16078055.2022.2146741
Popke, J. (2006). Geography and ethics: Everyday mediations through care and consumption. *Progress in Human Geography*, *30*(4), 504–512. doi:10.1191/0309132506ph622pr
Porananond, P., & King, V. T. (Eds.). (2014). *Rethinking asian tourism: Culture, encounters and local response*. Newcastle upon Tyne: Cambridge Scholars Pub.
Pritchard, A., Morgan, N., & Ateljevic, I. (2011). Hopeful tourism: A new transformative perspective. *Annals of Tourism Research*, *38*(3), 941–963.
Qi, H. (2020). Conceptualizing volunteering in tourism in China. *Tourism Management Perspectives*, *33*, 100618. doi:10.1016/j.tmp.2019.100618
Raymond, E. (2008). 'Make a difference!': The role of sending organizations in volunteer tourism. In K. D. Lyons & S. Wearing (Eds.), *Journeys of discovery in volunteer tourism: International case study perspectives* (pp. 48–60). CABI.
Raymond, E. M., & Hall, C. M. (2008). The development of cross-cultural (mis)understanding through volunteer tourism. *Journal of Sustainable Tourism*, *16*(5), 530–543. doi:10.1080/09669580802159610
Reay, D. (2004). 'It's all becoming a habitus': Beyond the habitual use of habitus in educational research. *British Journal of Sociology of Education*, *25*(4), 431–444. doi:10.1080/0142569042000236934
Sanderson, I. (2006). Complexity, 'practical rationality' and evidence-based policy making. *Policy & Politics*, *34*(1), 115–132. doi:10.1332/030557306775212188
Shamir, R. (2008). The age of responsibilization: On market-embedded morality. *Economy and Society*, *37*(1), 1–19. doi:10.1080/03085140701760833
Silk, J. (2000). Caring at a distance: (Im)partiality, moral motivation and the ethics of representation - Introduction. *Ethics, Place & Environment*, *3*(3), 303–309. doi:10.1080/713665900
Simpson, K. (2004). 'Doing development': The gap year, volunteer-tourists and a popular practice of development. *Journal of International Development*, *16*(5), 681–692. doi:10.1002/jid.1120
Simpson, K. (2005). Dropping out or signing up? The professionalisation of youth travel. *Antipode*, *37*(3), 447–469.
Sin, H. L. (2009). Volunteer tourism – "Involve me and I will learn"? *Annals of Tourism Research*, *36*(3), 480–501. doi:10.1016/j.annals.2009.03.001
Sin, H. L. (2010). Who are we responsible to? Locals' tales of volunteer tourism. *Geoforum*, *41*(6), 983–992.
Sin, H. L., Mostafanezhad, M., & Cheer, J. M. (2021). Tourism geographies in the 'Asian Century'. *Tourism Geographies*, *23*(4), 649–658.
Sin, H. L., Oakes, T., & Mostafanezhad, M. (2015). Traveling for a cause: Critical examinations of volunteer tourism and social justice. *Tourist Studies*, *15*(2), 119–131. doi:10.1177/1468797614563380
Singer, P. (2013). *A companion to ethics*. Oxford: Blackwell.
Slater, D. (1997). *Consumer culture and modernity*. Cambridge: Polity Press.
Smith, D. M. (1997). Geography and ethics: A moral turn? *Progress in Human Geography*, *21*(4), 583–590. doi:10.1191/030913297673492951
Smith, D. M. (1999). Geography, community, and morality. *Environment and Planning A*, *31*(1), 19–35.
Söderman, N., & Snead, S. (2008). Opening the gap: The motivation of gap year travellers to volunteer in Latin America. In K. D. Lyon & S. Wearing (Eds.), *Journeys of discovery in volunteer tourism: International case study perspectives* (pp. 118–129). Cambridge, MA: CABI Publishing.
Stafford, C. (Ed.) (2013). *Ordinary ethics in China*. London: Bloomsbury Academic.

Subrahmanyan, S., Stinerock, R., & Banbury, C. (2015). Ethical consumption: Uncovering personal meanings and negotiation strategies. *Geoforum*, *67*, 214–222. doi:10.1016/j.geoforum.2015.07.001

TaiwanICDF. (2004). *Volunteer Taiwan*. Retrieved from https://www.icdf.org.tw/web_pub/20040517150919Volunteer%20Taiwan.pdf

Taylor, E. W. (2008). Transformative learning theory. *New Directions for Adult and Continuing Education*, *2008*(119), 5–15. doi:10.1002/ace.301

Theerapappisit, P. (2009). Pro-poor ethnic tourism in the Mekong: A study of three approaches in Northern Thailand. *Asia Pacific Journal of Tourism Research*, *14*(2), 201–221.

Tomazos, K., & Butler, R. (2010). The volunteer tourist as 'hero'. *Current Issues in Tourism*, *13*(4), 363–380. doi:10.1080/13683500903038863

Tomazos, K., & Murdy, S. (2023). The limits of transformation: Exploring liminality and the volunteer tourists' limbo. *Tourism Recreation Research*, *49*(6), 1209–1221. doi:10.1080/02508281.2022.2162695

Tronto, J. C. (1993). *Moral boundaries: A political argument for an ethic of care*. New York: Routledge.

Tu, Y.-C. (2011). Volunteer tourism: What changes do we make?: Study on international volunteer tourists from Taiwan. Paper presented at the 29th EuroCHRIE Annual Conference "Tourism and Hospitality, Drivers of Transition", Croatia.

Turner, E. (2012). *Communitas: The anthropology of collective joy*. New York: Palgrave.

Turner, V. (1969). *The ritual process: Structure and anti-structure*. London: Transaction Publishers.

Varul, M. Z. (2009). Ethical selving in cultural contexts: Fairtrade consumption as an everyday ethical practice in the UK and Germany. *International Journal of Consumer Studies*, *33*(2), 183–189.

Vrasti, W. (2013). *Volunteer tourism in the global south: Giving back in neoliberal times*. Oxon: Routledge.

Walzer, M. (1994). *Thick and thin: Moral argument at home and abroad*. Notre Dame, IN: University of Notre Dame Pess.

Wang, J., & Brown, G. (2024). A critical analysis of volunteer tourism in post-socialist China. *Current Issues in Tourism*, 1–16. doi:10.1080/13683500.2024.2309143

Wearing, S. (2001). *Volunteer tourism: Experiences that make a difference*. Cambridge, MA: CABI Publishing.

Wearing, S., & McGehee, N. G. (2013). Volunteer tourism: A review. *Tourism Management*, *38*, 120–130. doi:10.1016/j.tourman.2013.03.002

Weaver, D. (2015). Volunteer tourism and beyond: Motivations and barriers to participation in protected area enhancement. *Journal of Sustainable Tourism*, *23*(5), 683–705. doi:10.1080/09669582.2014.992901

Wilk, R. (2001). Consuming morality. *Journal of Consumer Culture*, *1*(2), 245–260. doi:10.1177/146954050100100211

Yan, H. (2003). Neoliberal governmentality and neohumanism: Organizing Suzhi/value flow through labor recruitment networks. *Cultural Anthropology*, *18*(4), 493–523. doi:10.1525/can.2003.18.4.493

Yan, Y. (2013). The drive for success and the ethics of the striving individual. In C. Stafford (Ed.), *Ordinary ethics in China* (pp. 263–291). London: Bloomsbury Academic.

Yea, S. (2018). Helping from home: Singaporean youth volunteers with migrant-rights and human-trafficking NGO s in Singapore. *The Geographical Journal*, *184*(2), 169–178.

Zahra, A. (2010). Volunteer tourism as a life-changing experience. In A. Benson (Ed.), *Volunteer tourism: Theoretical frameworks and practical applications* (pp. 90–101). London: Routledge.

Zahra, A., & McGehee, N. G. (2013). Volunteer tourism: A host community capital perspective. *Annals of Tourism Research*, *42*, 22–45. doi:10.1016/j.annals.2013.01.008

2 Old and new ethics

The making of moral self

Moral landscape in Hong Kong and Taiwan

This chapter explores the moral landscape in Hong Kong and Taiwan in the context of volunteer tourism. I am going to examine how this form of tourism has become an appropriate choice by considering the complex ethical structure within these societies. I analyse the rationalities behind participation in volunteer tourism within a broader theoretical framework. In other words, I attempt to understand the meanings these motivations hold in their daily lives and how the decision to participate in volunteer tourism reflects these ethical orientations. Rather than focusing on the ethical choices available to individuals, this chapter explores the embedded framework of predispositions that shape these choices.

In Hong Kong and Taiwan, the value systems, which have been shaped by their colonial legacies, remain deeply influenced by the dominant philosophical Confucian tradition. Confucianism, a strong root of values in Chinese and other Asian societies (such as South Korea, Singapore, Vietnam and Japan) for thousands of years, is a traditional system of social and ethical philosophy as a way of governing or a way of living. It governs social life through institutions like family, school and state, where values are enacted through rituals. These rituals, both aesthetic and ethical, connect the present self to ancestral pasts and future generations, fostering a shared sense of humanity (Rosenlee, 2016). Confucian values emphasise human relationships, roles, and obligations that contribute to societal perfection, with *ren* 仁 (benevolence) at the heart of Confucian ethics (Ma, 2014). *Ren* is the source of all virtues, the intentional cultivation of *ren* is a form of spiritual practice (Rosemont, 2015), with rituals performed in a way that nurtures both inner character and moral maturation. Taking more roles and obligations helps to develop *ren*, the quality or character of the person. The concept of *junzi* 君子, or moral exemplar, further links this ethical cultivation to social interactions, as *junzi* helps others improve themselves through the cultivation of *ren* (Humphreys, 2017). In Confucianism, freedom is viewed 'as an achievement term, not a stative one, such that we can only begin to think of becoming truly free when we want to meet our responsibilities, when we want to help others…and enjoy being helped by others'

DOI: 10.4324/9781003413929-2

(Rosemont, 2015, pp. 106–107). This sense of personhood, developed through relationships with others, extends beyond individualism and links personal freedom with societal roles and obligations. It is in this context that values such as *guanxi* 關係 (social relationships) and *qing* 情 (sentiment) become significant, as they are closely tied to *ren* 仁 (benevolence), *yi* 義 (righteousness), and *li* 禮 (courtesy). Through the cultivation of these virtues, Confucian moral agents, or *junzi*, are expected to contribute to societal betterment. In the following, I am going to explore and understand how this cultivation of good character of people is actualised in the form of taking part in volunteer tourism.

Appealing discourses of sending organisations

The objective of this research is not to explore the marketing strategies of sending organisations in mobilising people for international volunteering, but through interpreting their promotional materials of the two sending organisations I analyse the prompts used by the sending organisations and why such language is appealing enough to make one to participate in volunteer tourism.

One of the posters TBW used was a ragged kid in a poor village narrating a food scarcity problem while ELIV depicted a shattered house in rainy season on its website. These visual representations conveyed the message that it is 'time for change' and volunteers were needed. The sending organisations were tapping on the motivations of volunteer tourists to make a change, which has been reported in the findings of many previous studies. The language of development embodied in the photos used tend to construct and reinforce the notion of 'Third World' and the imagined problems, similar to that used for gap year projects in the West in encouraging students to take part in development projects with a 'get on with it' attitude and assumed ability to do it (Simpson, 2004). However, what has remained unpacked is what these materials or messages really speak to? How do they speak to the ethical framework of these potential participants so that volunteer tourism becomes an appropriate choice?

Institutionalisation of service obligation in formal education

International volunteering or volunteer tourism programmes are commonly called service trips or overseas services *(海外服務)*. The notion of service has been extensively promoted in schools in Hong Kong and Taiwan. The idea is encapsulated in school mottos to convey the values, ethics and principles to bring inspirations in learning and even life. These mottos guide students to think about the kind of person they want to be and what they should be doing, thereby extending their actions beyond school. Many schools have been established and supported by religious bodies, mostly Christian and Buddhist, which incorporate religious ideas and values of cultivating characters such as compassion, empathy and kindness, as well as eliciting practices in the school mottos. While these mottos encourage the accumulation of good qualities in general, they also emphasise extending care to others. Helen from Hong Kong gave an example of her alma mater,

> our school motto is 'Omnia omnibus' which means 'All things to all people', that we should serve others, serve people around us. So, I have developed such concept or idea that in our daily life we do not only do things for ourselves, we also see if we can do anything to serve others.

'Service' was used more often than 'volunteering' or 'help' by the informants. Voluntary service is not new in Asia, and in Hong Kong and Taiwan in particular; but service learning has been introduced into the formal education relatively recently. It is defined as

> [a] course-based, credit bearing educational experience in which students (a) participate in an organized service activity that meets identified community needs, and (b) reflect on the service activity in such a way as to gain further understanding of course content, a broader appreciation of the discipline, and an enhanced sense of civic responsibility.
>
> (Bringle & Hatcher, 1995, p. 112)

Service learning has become a key experiential learning component of the new curriculum in Hong Kong and Taiwan. In Hong Kong, following the introduction of a new educational structure in 2009, Liberal Studies has been re-introduced (first appeared in the secondary school curriculum in 1988) as one of the four compulsory core subjects (Deng, 2009). This subject aims to

> help students understand themselves, and their relations with others and the environment in which they live. The intention is not to turn students into specialists in any well-defined academic field, but to enable them to become informed, rational and responsible citizens of the local, national and global community.
>
> (Hong Kong Curriculum Development Council (HKCDC), 2007, p. 4)

Practically, the intended learning outcomes of Liberal Studies are achieved by instrumentalising service learning or learning outside classroom. Other learning experiences (OLE) or experiential learning experiences such as moral and civic education, community service, and aesthetic and physical activities are emphasised to bridge the subject materials into practices. Service learning is the device or strategy of instilling a certain set of moral values which inform the act of volunteering as the rational choice. For example, Ona said that she was required to earn 'experience hours' under OLE when she was in secondary school, while Gloria had to fulfil compulsory service hours in summer vacation during her junior secondary school years. Similarly in Taiwan, while volunteerism in general has been promoted since the 1970s, a new 12-Year Curriculum for Basic Education has been reformed and implemented since 2014 based on the three ideas 'spontaneity, interaction, common good' (Chen & Fan, 2014). The idea of 'common good' encourages learning not just for oneself; students should not only care about themselves or their immediate circles but also the wider society and the world. Service obligation was a graduation requirement, it is required in some universities in Hong Kong (Ma & Chan, 2013); I personally experienced it. Xena from Taiwan also said,

> it's compulsory service in school. We had minimum service hours as a graduation requirement, all over universities in Taiwan, it's quite common and encouraged. I remember in Year 1, I forgot the theme of the activity, our lecturer took us to volunteer, we went to sweep the floor.

This sweeping floor was similar to what York, who is in his 60s, did in his earlier age. In the past, volunteering was largely service in nature, initiated by local community or local governmental offices. What makes activities like sweeping the floor different nowadays is that it has been institutionalised, as part of the formal education.

The idea of global citizenship is also coming through both education reforms. This theme is specified under the Module on Identity in the Hong Kong Liberal Studies curriculum framework (HKCDC, 2007), while it is not explicitly highlighted in the new curriculum in Taiwan[1]. Informed about their relations with others, human and non-human, the young generation has been encouraged to develop a sense of responsibility that is no longer limited to the local context but extends to a wider, global community and its challenges. This has helped to instil in the younger generation the idea of being a responsible global citizen, encouraging them to learn about other cultures and gain new experiences. As Helen said,

> people usually say, young people should have a global vision, so I want to understand different cultures and know what is happening in other parts of the world, that's the concept.

However, based on how this phrase – global citizenship – was used or talked about, its meaning and how to realise it did not seem to be clearly understood and in turn has become a buzzword sometimes. It seems that the nature of the activities was not the matter of concern; rather, it just became a requirement. Daphane said,

> we always hear about global citizenship, which makes you feel, ya, taking care of yourself is not enough, we shouldn't just think about ourselves.

Under the influence of global citizenship education, students have been conditioned to be more responsive to and thus responsible for narratives that calls for understanding and taking actions to address global issues. Here, volunteering is seen as a responsibility for graduation, which would mean that it is not based on ethical considerations or even an activity that is unquestioned by the students. Therefore, when Selina first saw the poster of the service trip she joined, she wanted to help those on the poster, as she said,

> I feel like we have been taught since secondary school that we should help others if possible, as we are all living in the same world, those concepts, global citizenship.

This speaks to the active dissemination of discourses of responsibility through formal school education instead of their personal motivations or initiatives (e.g. doing good things in this case). The students are mobilised when the narratives and practical resources or opportunities are made available to 'enable them to act as "responsible" subjects' (Barnett, Cloke, Clarke, & Malpass, 2010, p. 40). Their response was not necessarily about the broader social and environmental issues; rather, it was the act as the logical, unquestioned choice for being 'responsible subjects'. This is, as Shamir (2008) suggests, 'an interpellation which constructs and assumes a moral agency and certain dispositions to social action that necessarily follow' (p. 4). It has reconfigured the roles and identities of the students into global citizens (Beck, 1992; Kelly, 2001), attributing such qualities and statuses to these responsibilised agents. That is also a means of adding quality, in the *suzhi* discourse, to these young individuals. This raises the question of whether the accumulation of service hours would be conducive to the production and accumulation of moral capital, as this is to some extent determined by how the students perceive and receive such service hour accumulation mechanisms, and by other values at play.

Under the strategy of institutionalising service as a graduation requirement, the volunteers believed that a culture of compulsory volunteering has grown in schools; volunteering has become a collective action among peers. Strategies such as the point earning system and the service award scheme have created an atmosphere where students are very active in participating in service activities. Positively received, some students regarded it as a new experience instead of a service obligation. It is a unique experience in the sense that it is different from joining student societies or attending lectures; this relatively new form of extra-curriculum activity came as something they 'want to' rather than 'have to' do, shifting from an obligation to a norm or culture. Xena said,

> the several volunteering experiences with the lecturer have somehow made me know what volunteers do, start to learn more about this field, and want to get more involved.

After sweeping the floor, Xena had more volunteering experiences initiated by the same lecturer. Volunteering here does not necessarily become a discrete choice of doing good made by an autonomous individual; it could simply be an instrumentalised appropriate choice. In other words, the active participation or positive attitude of the students could possibly be an unconscious response to the culture, the requirements or the availability of such new option in schools, instead of the need for volunteers. These positive responses to the strategies have helped to inculcate dispositions, instilling a notion of responsibility. Many of them said that they want to become 'a person who can help others' through their volunteering experiences in school and university. A sense of empowerment is running through, making them a category of moral subjects that 'they can' rather than simply 'they should'. Embedded in this shift from service obligation to a norm or culture is the cultivation of their inner character and ethical maturation, or their *ren*, instrumentalised in formal educational institutions. Taking up more responsibility and helping others guides them along the pathway to developing the moral self.

On the other hand, not all the informants responded positively to these strategies of instilling moral values when they were students. They treated the graduation requirement as a duty or a task instead of developing it into an interest or a habit, like Selina said she 'just wanted to get it done' when she was in school. When the students challenged what was believed to be appropriate or reasonable, this would then interrupt the production, accumulation and reproduction of moral capital (Davey, 2012). It did not become a preferable or moral action but a tick-the-box kind of thing they needed to fulfil. It did not facilitate internalising the disposition such that the students genuinely wanted to help others. Rather, the choice has been internalised as the thing that they 'have to' do. This is different from the moral selving through which the individuals actively develop themselves to become more virtuous. It is also different from responsibilisation strategies in the West that lead to an interpellation making volunteering an unquestioned choice. The students have become institutionalised moral subjects instead of responsibilised moral subjects through the act of volunteering. It is an instrument in school, a structure, through which the conduct or the behaviour is determined. In the meantime, the students have the agency not to perceive volunteering as the choice to take in the process of doing it involuntarily given that it is the activity 'to meet a target number of service hours', as Selina said, when the technologies of the self are exercised within the structures of coercion or domination (Foucault 1988). This is evident from some volunteering programmes designated for high school students in Taiwan for this purpose. Although some students did not respond to these strategies positively, the culture of compulsory volunteering has been internalised and ingrained; the dispositions have added another layer to the habitus and they might take moral actions subsequently in modified ways reproduced from the dispositions.

This is the case of service-learning but different from the gap-year culture in the West. Gap year is no longer about dropping out but rather travelling as a global citizen. Simpson (2004) has also noticed that gap year is increasingly getting incorporated into formal educational and employment structures and institutions by being tied to formal education and better access to employment with a language of such as 'success' and 'graduation', meanwhile autonomy is still embedded in this decision. On the contrary, it is more than a 'something for the CV' thing in the case of Hong Kong and Taiwan as it does not emphasise on skill development (Lyons, Hanley, Wearing, & Neil, 2012). It is rather largely framed by the service obligation and institutionalised culture of volunteering, in which obligation has somehow been internalised into a norm like the blood donation in China (Jun, 2011). The global citizenship constructed from this institutionalised culture of volunteering with a focus to develop a moral self entails both ethical citizenship and dutiful citizenship (Cloke, Johnsen, & May, 2007), distinguished between the 'want to' and 'obliged to' motivations.

A 'moral self' in different generations

In Hong Kong and Taiwan, family is usually the first place of learning and is influenced by Confucian thinking. One of the Confucian values that is highly relevant to human relationship, originating in the family, is the idea of relationality – the starting point and foundation of our ontological existence throughout the lifetime (Rosenlee, 2016). As Rosemont (2015) points out, 'for the Confucians there are only interrelated persons, no individual selves' (p. 93). From this, it is irrelevant to consider a person as either selfless or altruistic in Confucianism, when the individual bears some roles and obligations, and living those roles and obligations in relation to others is to produce their existential personhood. Although *ren* – the core value of Confucianism – tells the importance of benevolence, everyday social interactions are guided by differentiated ethics. Helping others reflected the ordinary ethics of care especially in agricultural communities in the past when families were giving hands to harvesting or looking after kids in the neighbourhood. Such ethics are still in play in informants' families especially when they observed and experienced how their parents treated the proximate such as immediate family members and relatives compared to the distant others. For example, both Charlotte and Xena explained that in their families, relatives would help each other if needed but there was no such concept of helping 'those you do not know'. People care for others based on their relations, the ties between people. Ethan's story of how his grandmother treated her domesticated animals and neighbours further explained such differentiated ethics. He said,

> My grandma told me many bizarre stories, about how she took care of her cows… a chicken she had kept for 10 years, that's her first chicken, it ,,, brought her a lot of fun, so she kept it until it died… My grandma was quite compassionate… when an old man in another village passed away, she was reminiscing the old days she had with him in the farm… full of emotions.

As we can see here, compassion is framed with these ordinary ethics, their relations with the 'others', the *guanxi* (social relationship) and *qing* (sentiment), rather than a logical framework based on justice. The connection with others and the emotion induced is based on the strong sense of concentric realms of care. This particularistic ethics of care is acceptable and reasonable in Confucian families as care should be extended from the immediate family to relatives, then community and the state when the roles and obligations are fulfilled from the core. The fact that Ethan found the stories bizarre would suggest that this Confucian focal value is not strong among the younger generation as it is in the older generation. These traditional values are part of the school curriculum. When the recent curriculum reforms in Hong Kong and Taiwan both introduced the concept of global citizenship and reiterated service learning as the pedagogical approach, the conceptualisation of care and responsibility has been changed in terms of the scale and radius. This has, to some extent, changed younger generation's perceptions of responsibility and thus geography of compassion. When concepts such as global citizenship and service learning entered the curriculum of the younger generation, they had more effects on the habitus of the younger generation than on the older generation, whose values were already so deeply embedded in their habitus that they were less likely to change. It was more prominent when active strategies were implemented to instrumentalise service obligation in schools while traditional Confucian values were passive and implicit 'rules of the game', in Bourdieu's terms (Lamaison & Bourdieu, 1986), the basis of structuring the experience has been restructured. New pattern of thoughts has emerged from the implementation of the new curriculum and the formalisation of services. This could help explain why, interestingly, most young volunteers did not flag up this particularistic ethics of care among their generation.

This differentiated ethics was particularly regarded as 'old ethics' of the previous generation by the volunteers. Camillia, for example, said her parents are more traditional, 'no extra time to think about others, look after yourself first'. This goes against the younger generation's value of looking after more than just themselves. Ona, who was 20 years old, also shared,

> my mum always thinks we should help as long as we can, but her value is more for helping friends instead of society.... But I think I am different from her generation that I look at a broader context rather than just people with closer relationship.

The younger generation is more influenced by the school education that promotes a global vision and active participation in helping to resolve global issues. This shows how the interplay of learning in school and family settings as well as the active strategy of formalising service obligation affects the habitus. Here, a new ethics of care and helping others is found more distinctively among the young people. 'Others' is perceived differently between generations; no longer being limited to the 'proximate others', they have started to extend their care beyond their circles of relations without attending to the ethics such as fulfilling obligations to parents before those in the outer concentric spheres and even strangers.

Traditionally, the ethics of filiality is practised through obedience and obligation. Children are expected to follow parents' arrangements and expectations and to take good care of their parents at their older age. A slight inter-generational tension may emerge when the children want to be independent from their parents to enjoy the freedom of doing something different. Amelie's experience showed a more direct and explicit conflict of values between generations. Feeling frustrated in getting her mother to understand her participation in service trips, she explained,

> I told my mum, I was planning to go to Cambodia, to build toilet, she did not bother to respond. She asked, 'why going there? Why not travelling with me instead?' Because I promised to travel with her after I became a teacher, I had not fulfilled that promise, so I put off joining the Cambodia project… because of my mum, I decided to travel with her in 2015 summer.

Over time, the desire for going on a service trip had not yet been fading out despite her mother's disapproval, which had driven her to empower herself through earning the programme fee, to negotiate and silently confront her mother with an opposing value. Meanwhile, in this field governed with Confucianism, the rules are still in place to be obeyed (Lamaison & Bourdieu, 1986). As Confucianism has been historically and culturally rooted as a way of governing and a way of living, people could not escape this rule of the game although not everyone is devoted to or actively enacts its values. Amelie still needs to fulfil her duties and obligations at home as a good daughter before taking up more roles and obligations outside the home. The moral relations in family, which is the centre of relations, are still being governed by Confucianism and thus fulfilment of assumed responsibilities is still emphasised. Here, it is not about the conflict between travelling for good cause or fun. It is rather about the contradiction of particularistic ethics of care, the ethics of filial duties and helping the distant others between generations. This is further demonstrated by York, in his early 60s, who explained how he could enjoy this 'freedom' and have the ability to extend the care to the distant others. He said,

> the primary condition for helping others is to make sure yourself are well-managed so that you have the ability to help others… Whether it is right or wrong, I think it is, you can extend the care and help to others when you have the ability. When you are not doing well but helping others, that's strange too.

He went on service trips because, to a certain extent, he had fulfilled his obligations for raising his children. Generally speaking, individuals have the freedom and choice to maintain or withdraw a certain role once they reach the age of reason such as adulthood. However, this is not the same in a Confucian perspective when the 'self' is always part of a social relationship (Rosenlee, 2016). Although the younger generation is enjoying and advocating for more freedom and autonomy and at the same time feeling more responsibilities from the institutionalised notion of service obligations and global citizenship, they are still bound by a limit of possibilities. This explains that habitus has enabled the production of different forms of practices that are unpredictable but still reasonable falling within the limits of the regularities (Bourdieu, 1990).

Although parents' way of doing good may be challenged by their children who have acquired another layer of dispositions into the habitus outside home, the values of moral selving are compatible and internalised for developing a person with good qualities. This depends on how the children's patterns of thought are challenged or how they respond and negotiate. When talking about the influence of her parents on her participation in volunteering, Wendy was not aware of it or did not want to admit it until different narratives were linked up. She said,

> we are very traditional Confucian family, when I was small, my dad told me that he wanted me to become a person like Confucius, those kinds of people whose influences are still existing after 2000 years. He said … can you use your actions or knowledge to influence others? … I was like, rolling my eyes. He used to have such expectations on me, to have some sort of influences. He does this from time to time, as he likes reading books about philosophy, likes talking about principles and values, andI listen by the side.

Although she did not entirely agree with what her father said, the dispositions were unconsciously inculcated and accumulated. She interpreted and practiced in her own way of being a good person making some contributions. Volunteering was her way of expanding influences, and she believed she has gradually become a person who can create some influences. She continued,

> it's a value, in my mind, maybe it influenced me to make the decision [of volunteering], but I don't have much consciousness… but my dad's influence on me is quite great in general, it's like something implanted in my mind without self-consciousness. I need to think how to expand my influences, so I did volunteering, he's quite successful although I don't want to admit that … When he tries to share his thoughts on life, I am like 'okay, fine, no need to say that much, I got it, understood'. Sometimes, if I agree I really listen, I think, right, I would think how I can do it if I agree and want to do it.

Instead of mobilising the cultivated moral subjects for actions, these values are not circulated explicitly. As many volunteers said, these values are more embodied in practices in the ordinary aspects of everyday family life. York, for example, has engaged his children in volunteering by making it an ordinary family activity. He joined local service activities such as tree planting with his children when they were small. According to him, his elder daughter has developed an interest in volunteering since then and went on a service trip to Cambodia a few months earlier than him. By engaging the children in the practices of helping others rather than only telling them what is good and right to do, the transmission of moral values is tacit but more direct. Normalising the practice of helping others as part of the family activity, part of the daily life, facilitates the unconscious inculcation of habitus. The transmission and translation of values into habitus and cultural and moral capital lies on the cultivation of the qualities of the person, *ren* as the core value and other associated values, and roles and obligations of the self in relation to others. This work of self-forming has helped to commit the children to certain ends – to become a virtuous person who bears the quality of acting morally.

Identifying role models in school

Many times, the ethics of care, responsibility and doing good are tacit and implicit. Practising these ethics and accumulating qualities through such practices is not a straightforward teach-and-learn process. Besides formalising volunteering as the technology or instrument, exemplarity is another strategy in school. Identifying the moral exemplars through ascribing an exemplar status to some students who are enthusiastically participating in volunteering helps to demonstrate and define the quality of a good person, which may be effective in promoting and facilitating self-cultivation and accumulation of *suzhi* (Cody, 2018; Matsuba & Walker, 2005). For instance, Ilsa has been regarded as the 'kind of person doing it' following a sharing on her volunteering experiences in school. After a visit to a charity organisation when she was in secondary school, Ilsa had a lot of reflections and interests and later signed up as a volunteer for that organisation in order to know more about the issues she was concerned about such as poverty and human trafficking. Being noticed by her teacher, she was invited to share her experiences at the school assembly. As her talk was well-received, her peers started to position her as the right person to ask whenever volunteering activities came up, while she also started to develop a felt responsibility of volunteering. She said,

> people thought I did it very well, so I was positioned as doing something like that. Since then, people would ask me to help as they thought it's my 'field'. From time to time, people are saying it, then I thought, oh, that's what I have been doing. So, when I saw the promotion of the trip last year, I felt like it's my stuff, I should do it, had some sense of mission, after being confined to this image, not sure if it's good or bad, but with this image, I would keep doing towards that direction.

This, on the one hand, is the strategy in school to organise visits to charity organisations hoping to promote the idea of volunteering and through which instil some moral values on students. Ilsa was one of those who came back with the outcome that the school teachers wanted to see – the kind of individual or role model produced through the strategies was then displayed to further encourage students to take on more responsibilities and become a better person. Her story acted as a successful example of bridging between doing something good and becoming a better person, forming a role model to be learnt from (Matsuba & Walker, 2005). On the other hand, this exemplarity created the sense of mission which, to a certain extent, helped sustain the practice of volunteering in the long term as the way to maintain the status as a role model depicting its qualities. It has reinforced the habitus of the role model and thus volunteering has been rationalised as the appropriate choice. It then formed the correlation that it became the sort of things they should do as they fell into the category of moral subjects. Therefore, when she saw the promotional poster, she felt she should do it as it came to her as an unquestioned choice.

Such strategy of identifying and displaying moral exemplars is also evident in Taiwan's vision in education. Part of the Youth Development Affairs' vision is to 'empower volunteer service competence; promote diverse youth volunteer service, subsidize youth for forming teams to organize all kinds of services; organize national competitions for excellent youth volunteer teams and conduct awards ceremonies to reward model volunteers' (Ministry of Education, 2017). In order to identify and reward model volunteers, schools have set up point award system or award ceremonies to encourage students to participate in volunteering and earn points for service hours. For example, Hestia shared that when she was at the university, there was a ranking among the cohort for their accumulated service hours and the top three received an award. She explained:

> before going to the university, I did not know there was a mechanism for accumulating service hours. Our faculty's graduation requirement was 100 hours, other faculties only required 24 hours. At first, 100 hours sounded a lot to me, but I graduated with 1200 hours. But 1200 hours was just the 4th in our class. I was not sure about the highest number, but at that time we could apply for a service award based on the number of service hours, I was not on the top three. I also joined student societies on top of doing service. Although the hours spent for organising events were not counted, I was given the award. Someone said it would have been strange if I had not got the award with 1200 service hours.

When the act of doing good is evaluated or even subjected to a competition, it becomes an act of signalling, although it may not be the intention of the volunteers. Setting up role models was aimed to strengthen youth volunteer participation and development. However, it resulted in a focus lying on the number of hours and award. In this sense, it means that quality can be quantified. The exemplarity may help sustain the practice of doing good, which may also be because of the felt responsibility. In this example, it does not seem to help internalise the moral values through the role models; it may then be reduced to an exercise of fulfilling the graduation requirement. It could be seen as a strategy used in school to add quality to the students not only responsible for wider social issues but also more governable in school. Having said that, we could not deny the possibility or potential of such mechanism in creating the trend of doing good and subsequently producing instrumentalised moral subjects, because Ilsa and Hestia, for example, have become long-term volunteers.

Crafting a class of moral subjects between old and new ethics

What is emerging from above is the interplay of the service obligation, global citizenship, new and old ethics, and thus possibility of extending care, cultivation of *ren* and other Confucian values. This chapter has unravelled the social attributes or cultural setting in which these personal qualities are made sense of or defined. All these have reflected how the structures of the social world are incorporated in the body of an individual in certain ways that structure how the individual acts, at the same time producing a notion of quality as a form of moral self.

Instead of reflecting the habitus of a class, it is about the attempts and outcome of crafting a class of instrumentalised moral subjects. The moral self in this class is not simply a responsible, rational, autonomous neoliberal subject with choices; the moral subject aims to become a good person, a better self through the instrumentalised strategies interacting with tacit traditional values. This is slightly different from the Western responsibilisation in the way that in this East Asian context, self-development or becoming a better self is the choice, and the outcome of the choice is not being ethical. This has shown a culture-specific form of technologies of the self. Thus, it is not about the choice of doing good or doing bad, or even doing nothing. This offers another lens to move the understanding of responsibilisation from positing individuals to make discrete choices, to looking at the overarching, complex ethical terrain in which individuals choose to do things from a particular instilled position.

When sending organisations prompt to the need to develop a virtuous self and the felt responsibility from a conjuncture created by service learning initiatives and new ethics of care and responsibility, the discourses of sending organisations become so appealing that volunteer tourism thrives as the appropriate choice particularly when it has emerged as a moralised, alternative form of travel. It is then important to unpack what 'change' in 'time for change' would mean for the self-making.

Note

1 A few Taiwanese informants said the phrase 'global citizen' or 世界公民 in Chinese has not yet been widely used in school but started to be circulated by NGOs.

References

Barnett, C., Cloke, P., Clarke, N., & Malpass, A. (Eds.). (2010). *Globalizing responsibility: The political rationalities of ethical consumption*. Chichester, West Sussex: Wiley-Blackwell.

Beck, U. (1992). *Risk society: Towards a new modernity* (Vol. 17). London: Sage.

Bourdieu, P. (1990). *The logic of practice*. Stanford, CA: Stanford University Press.

Bringle, R. G., & Hatcher, J. A. (1995). A service-learning curriculum for faculty. *Michigan Journal of Community Service Learning*, *2*, 112–122. Retrieved from http://hdl.handle.net/2027/spo.3239521.0002.111

Chen, H., & Fan, H.-H. (2014). Education in Taiwan: The vision and goals of the 12-year curriculum. Retrieved 25 May 2018 https://www.brookings.edu/opinions/education-in-taiwan-the-vision-and-goals-of-the-12-year-curriculum/

Cloke, P., Johnsen, S., & May, J. (2007). Ethical citizenship? Volunteers and the ethics of providing services for homeless people. *Geoforum*, *38*(6), 1089–1101. doi:10.1016/j.geoforum.2006.07.005

Cody, S. (2018). Borrowing from the rural to help the urban: Organic farming exemplars in postsocialist China. *The Asia Pacific Journal of Anthropology*, *19*(1), 72–89. doi:10.1080/14442213.2017.1394362

Davey, G. (2012). Using Bourdieu's concept of doxa to illuminate classed practices in an English fee-paying school. *British Journal of Sociology of Education*, *33*(4), 507–525. doi:10.1080/01425692.2012.662823

Deng, Z. (2009). The formation of a school subject and the nature of curriculum content: An analysis of liberal studies in Hong Kong. *Journal of Curriculum Studies*, *41*(5), 585–604. doi:10.1080/00220270902767311

Foucault, M. (1988). Technologies of the self. In H. G. Luther, H. Martin, Patrick H. Hutton (Ed.), *Technologies of the self: A seminar with Michel Foucault* (pp. 16–49). Massachusetts: University of Massachusetts Press.

Hong Kong Curriculum Development Council (HKCDC). (2007). *Liberal studies curriculum and assessment guide (secondary 4–6)*. Hong Kong. Retrieved from https://ls.edb.hkedcity.net/file/C_and_A_guide/201511/LS_CAGuide_e_2015.pdf

Humphreys, J. (2017, 7 July). What does the rise of China mean for western values? *Irish Times*. Retrieved from https://www.irishtimes.com/culture/what-does-the-rise-of-china-mean-for-western-values-1.3276600

Jun, J. (2011). From commodity of death to gift of life. In A. Kleinman, Y. Yan, & J. Jun (Eds.), *Deep China: The moral life of the person* (pp. 78–105). Berkely: University of California Press.

Kelly, P. (2001). Youth at risk: Processes of individualisation and responsibilisation in the risk society. *Discourse: Studies in the Cultural Politics of Education*, *22*(1), 23–33. doi:10.1080/01596300120039731

Lamaison, P., & Bourdieu, P. (1986). From rules to strategies: An interview with Pierre Bourdieu. *Cultural Anthropology*, *1*(1), 110–120.

Lyons, K., Hanley, J., Wearing, S., & Neil, J. (2012). Gap year volunteer tourism: Myths of global citizenship? *Annals of Tourism Research*, *39*(1), 361–378. doi:10.1016/j.annals.2011.04.016

Ma, C., & Chan, A. (2013). A Hong Kong university first: Establishing service-learning as an academic credit-bearing subject. *Gateways: International Journal of Community Research and Engagement*, *6*, 178–198.

Ma, L. (2014). The comparison of Confucian Benevolence and Christian Love. *Studies in Literature and Language*, *9*(1), 145.

Matsuba, M. K., & Walker, L. J. (2005). Young adult moral exemplars: The making of self through stories. *Journal of Research on Adolescence*, *15*(3), 275–297. doi:10.1111/j.1532-7795.2005.00097.x

Ministry of Education. (2017). *Education in Taiwan 2017/2018*. Taiwan: Ministry of Education Retrieved from https://english.moe.gov.tw/cp-16-17192-F34D2-1.html

Rosemont, H. (2015). *Against individualism: A Confucian rethinking of the foundations of morality, politics, family, and religion*. Lanham: Lexington Books.

Rosenlee, L.-H. L. (2016). Against individualism: A confucian rethinking of the foundations of morality, politics, family, and religion by Henry Rosemont Jr. *Notre Dame Philosophical Reviews*. Retrieved from https://ndpr.nd.edu/news/against-individualism-a-confucian-rethinking-of-the-foundations-of-morality-politics-family-and-religion/

Shamir, R. (2008). The age of responsibilization: On market-embedded morality. *Economy and Society*, *37*(1), 1–19. doi:10.1080/03085140701760833

Simpson, K. (2004). 'Doing development': The gap year, volunteer-tourists and a popular practice of development. *Journal of International Development*, *16*(5), 681–692. doi:10.1002/jid.1120

3 Grounding morality

The being of the moral self in moral community

On the front page of ELIV's website about its vision and objectives, a section with a heading of '*影響力哲學*' (literal translation as 'a philosophy of impact') says:

> As people change, the world begins to change. The world is undergoing dramatic changes and crises, and we have all become aware of the need to take a serious look at the future of the planet and humanity. We believe that the best version of each individual gathered in a community will be the best version of the community, and will naturally apply the most appropriate ways of treating each other and the planet.
>
> (ELIV, 2018, translated)

While the potential impacts, both positive and negative, on the host community and volunteers have been discussed and acknowledged in the literature, this vision of an NGO practitioner leads us to think how we could conceptualise 'impact' of volunteer tourism through forming a community out of the volunteering space. Taylor (1991) has suggested that disenchantment of the world and loss of community are caused by individualism and alienation while Arai and Pedlar (2003) have seen moral loss and social crisis of alienation as emerging social problems. In this chapter, I set out to explore the kinds of relationship that foster certain kinds of interactions to fill a perceived moral vacuum and help to unveil the best version of the self.

I propose that moral communities are constructed by the volunteers through their practices of doing belonging and in so doing making themselves accepted into the moral communities for moral fulfilment. At the same time, liminality allows the volunteers to do what is otherwise restrained in ordinary life. These moral communities appear as some other moral worlds at temporal spaces or different timescales. This chapter then aims to discuss how ethics of care, responsibility and obligation are enacted in a liminal tourism space by looking at different 'we' relations. This helps us to understand how the moral self is cultivated and experienced in relation to others. I have identified three forms of moral communities – moral community of volunteers, moral community of like-minded people and moral community with the locals. In terms of the

DOI: 10.4324/9781003413929-3

nature of relationship, I borrow Heidegger's notions of 'being-with', 'being-alongside' and 'being-in' to understand the modes of being enacted by volunteers in these moral communities.

For Heidegger (1953), 'Being-with' (*Mitsein*) is an essential part of 'Being-in-the-world' (*Dasein*), as he explains,

> Being-with is such that the disclosedness of the Dasein-with of Others belongs to it; this means that because Dasein's Being is Being-with, its understanding of Being already implies the understanding of Others'.
>
> (p. 160)

'Being-with' emphasises relationality, as Heidegger (1953) suggests, for being is always relational and shared. He emphasises that engaged being-with is always already 'being with others'. 'Others' here does not mean anyone else but me. In his view, individual *Dasein* (a human being is not reducible to a subject) is not considered first as a given and then used to construct its relationship with the surrounding world, but is defined through its relation to others, to things, and to oneself. I borrow this term to refer to the ways volunteer tourists engage with others and in close relationships, especially when moral codes are suspended or through searching and revealing the different dimensions of the self.

As of 'being-alongside', Heidegger (1953) states that two 'things' cannot 'touch' each other because there is always a minute space between two 'things', even though this space is perceived as nothing. When applying this to look at two entities, this 'touch' implies human closeness or proximity. The state of 'being-alongside' also stresses a partial connection that involves attachment and detachment (Latimer, 2013). Thus, 'being-alongside' is used to understand the relationship that arises from working together, which involves responsibility in a collective setting, or collectivity.

'Being-in' is usually associated with involvement in a context or situation. In addition to 'inness', 'being-in' can also be used as a way to describe patterns of existence (Heidegger, 1953). Here, I borrow this term to refer not only to a spatial relation, but also to the state of feeling and doing belonging. This allows us to understand how the states of 'being-with' and 'being-alongside' help to produce the state of 'being-in', or how active doing of belonging fosters 'being-with' and 'being-alongside'.

Moral community of volunteers

Belonging to a moral community of volunteers has given them an intense experience of being in human relationships that they thought had long been lost. The coming together of people as a group to volunteer creates a web of relationships which allow the volunteers to feel the sense of collectivity and care. Bowring (1997) has argued that the moral fabric of the society has been disintegrating, while Arai and Pedlar (2003) have suggested that society is 'a meeting place for individual wills' (p. 195). Many informants raised similar concerns about the individualistic culture in Hong Kong and Taiwan. Their concerns included how people seek self-realisation and their own goals at the expense of others, the competition resulting from neoliberal responsibilisation, which have further divided the society, leading to a loss of social cohesion and different social problems. They expressed that people tend to avoid responsibility and commitment for the common good, comparatively their volunteer companions are very willing to give and even sacrifice as well as to express their care and empathy. Feeling frustrated and disappointed at the lacking sense of obligation and the loose ties between people in society, they believed that the sense of collectivity running through the community of volunteers helped to reconnect people by actualising the meaning of obligation, responsibility and care in various forms, and thus moral fulfilment beyond the everyday context, which is essential to maintain the functioning of society. Dawn gave a more nuanced account for this notion of community:

> In many times, especially when we start working, we will gradually forget this thing, but what we need is mutual care, caring for others. I think that's something to maintain the function of this society so that it won't turn too terrible. I think it's like a network, I imagine it like a spider web. Each of us represents a node, but we can't survive on our own node, so we have to connect with other nodes to form a line, and lines are woven into networks, I think society should be like this, also society is indeed formed in this way, but from time to time, people would just forget. Volunteering helps us find back the ability that human beings possess, everybody has this ability.

Dawn highlighted the relationality beyond simply being related to others but 'a state of humanness' (Neville, 2008, p. 251). What they were looking for was this sense of interdependence and reciprocated care among people like a spider web. Here, we could see the quest for being-in a community and recognition for being part of the community, and for the interactions and reciprocity in the relationships. Dawn described the best form of our society and that people have the ability to maintain such web of relationships. This leads us to think whether or how the form of social relations has been shaped by the interplay of the old and new ethics (see Chapter 2) in the current individualistic society. A moral community of volunteers is formed through volunteer tourism, which creates the space to regain this lost sense of mutual construction between self and society.

The formation and development of this moral community are not unidirectional or one-dimensional as it is an organised group of travellers rather than simply a collection of individual tourists. The two organisations had a screening process to select applicants with the qualities needed to help achieve the project's goals. This process thus created the post-place community of volunteers who joined the community for seeking some sort of moral fulfilment. A certain sense of moral relationship was produced through their experiences of collectivity and the embodied fulfilment from the doing of good.

Being a member of this moral community, volunteers had to engage in the collective in order to be recognised as part of it. It was assumed that everyone shared the tasks, no one should, or would, avoid the responsibility, the responsibility of being part of the group. At the same time, most of the informants believed that people who join service trips mostly share a common goal and are willing to give. With this sense of collectivity and responsibility, as one of the team leaders said, the volunteers 'worked at their utmost, no need to allocate'. Volunteers were moving flexibly, showing up at the right place for different tasks and coordinating very efficiently, guided by the assumed responsibility and facilitated by their rapport and team work. Xena gave an example,

> we made the division of labour instantly, very quickly divided into small teams, such as when we were asked to line up for a relay, we immediately formed a line, or when we needed six people for moving the concrete then six people appeared there instantly, or someone went up to help when someone else said it's too heavy.

The work seems to be simple and straightforward although physically demanding, what is more important here is the group endeavour. As a collective action, each volunteer was one node forming the relay line in the sense that the group was helping and interacting with one another, actualising the willingness to give and working towards the common good which the volunteers wanted to see and experience. Despite the challenges of doing things that they had never done before, such as construction work, the volunteers stressed the importance of everyone taking part in the activities. For example, TBW Team-A was tasked to build a footpath along the river by driving the logs into the riverbed. The village head who was guiding the task did not expect them to finish in one day, and the team did not manage to accomplish it either, but each of them had tried plunging the logs into the mud while the others gathered around to give support. Similarly, the ELIV Team also got involved in every single step of building the house and toilet. Even when lifting up the structure of the house, everyone in the group was trying their best to have their hands on it. This embodied the collective action, that each one of the team was part of this community. Undeniably, it was about doing good; but more importantly, it was about the common good as a result of the group endeavour that was meaningful to them (Etzioni, 1995).

Attempts to show their presence in the team work were equally important if they could not participate physically. Adam and Quinn were treated differently by their team members due to their different efforts of doing belonging. From the very beginning, Adam of TBW Team-A was treated as an outsider because the group did not see enough involvement and commitment from him in all tasks. Their first tasks in the fishing village were to collect mangrove seeds in the morning and work in the seedling nursery in the afternoon. The work included getting mud from the river, and filling the pots with the mud and seeds collected for nursery. Most team members commented on Adam's reluctance to get his hands and feet in the mud to help while the rest of the team and villagers worked towards a common goal. Charlotte was one of them, saying,

> if you joined a service trip, you should have expected the worst, expected that the environment was not that pleasant. We were there to plant mangrove, in the sea and river, I thought these should have been expected, but he had not.

Adam should have expected such working environment. In addition, he should have expected collective efforts towards a shared goal that gave a meaning to the project team, as this was the solidarity and norm of this moral community that they believed sustained the practical connection among volunteers from the beginning of the trip. Meanwhile, Adam did not deny this reluctance. He got several cuts on his legs when collecting mangrove seeds on the first day, as did other volunteers. Adam was afraid of getting an infection from working in the mud as the river did not seem clean to him. After 'I sensibly analysed and thought he's [Ethan's] right', Adam said, he was sitting in the reading corner doing nothing because he did not want to go down to the mud. This shows the struggle between the rationality of not getting infected and involvement in a collective action, at the same time the guilt of not doing it or not doing enough for the belonging and recognition from the group, which he said he regretted afterwards.

Similarly, Quinn of TBW Team-B injured his foot while riding a motorbike on the third day in the village. Although his wound could not come in contact with water and limited his mobility, he was trying to participate and help by mixing the concrete at the construction task. Instead of blaming him for being injured and unable to work, his team members did not allow him to go to the site to work, telling him to 'go out to sit, do not move'. Despite this, he was still trying to help by regularly handing them drinking water at the site and by building a few bricks of the wall.

The different treatment by their respective team members explains the importance of their efforts to show that they participated in the collective action and did what was expected of them as a group after they decided to go on the service trip and got accepted into this community. Adam's behaviour was seen as different because everyone in the team got cuts on their hands and legs, but he was the only one who cared about infection, did not participate in the subsequent tasks or stay around the nursey to help from the footbridge as two other volunteers had done. Here, it is not the debate on getting infected or accomplishing a task; it is rather on the ethics of being-alongside and attempts of being in this moral community. On the contrary, Quinn maintained his part in the team to show a shared sense of responsibility. He showed his presence and involvement, living up to the assumed qualities of this community. Besides, what Charlotte said was not just about the harsh conditions but also participation and integration of the volunteers in accomplishing the collective goal. Adam felt regretted and frustrated because he was absent from the experience, as what constructed the experience and inscribed the memory of his participation was physically being there, climbing up the trees to collect mangrove seeds and working in the mud for the seedling nursery. The footsteps in the mud embodied his presence and the work he was doing together with the group. Nevertheless, he was not doing the same thing as others at the same tempo and rhythm. On the other hand, even though Quinn mainly helped to mix the concrete and later looked after others, his efforts were embodied through the concrete he mixed and the bricks he laid in building the house as an outcome of the collectivity. Adam and Quinn demonstrated different levels of attachment to (or detachment from) the collective action and the team. This then resulted in different intensities of their being-alongside and thus being-in or recognition by the group.

This coevality is determined and shaped by how ordinary ethics is played out in this moral setting. Going on service trip is usually associated with harsh environment which are not much experienced in volunteers' home country. Being in the group, volunteers should have the expectations of living in the local way, or the reasonable way of living up to this 'membership'. For instance, Amelie said,

> I am quite picky in Taiwan, but when I am in another place, I just accept how it is, like in Qinghai we peed in the nature, there was no toilet, what can you do, still care about cleanliness? Everyone was doing it. In Cambodia, roads and streets are sandy, sand goes into my shoes, I just clean my feet at night, just get dirty, I am fine with how it is, that's their culture, that's Cambodia.

Everyone was doing it. Doing it did not only show that they were doing the right thing in that setting, but it also meant that they followed the values or norms of the group, which made them feel integrated. Feeling disintegrated could be as harsh as the living conditions there. This is because taking up challenges and self-fulfilment both were their purposes of going on the service trip; the former bears the transformative potential (discussed in the next chapter) while the latter can be achieved through being part of this moral community. Because of this ethics of being-alongside and being-with, volunteers were committed to the work without complaining about the hard work and harsh conditions such as the high temperature, or internalising any struggles. For instance, Felicia found the heat unbearable at one point when they were building the house but she thought it would not be appropriate if she moved to the shade. She shared,

> people never complained, although it was so hot, building the house in the sun… only me complaining secretly… about the hot weather, why couldn't we go to the shades? Once when we were cutting the bamboo sticks, there was a shade next to us, but nobody moved to the shade. I actually wanted to move to the shade, but since nobody moved so I did

> not dare to do so. It was very hot, the first day of building the house. At that time, I was wondering why they did not ask us to stay in the shade, I was thinking to myself, but seems nobody had noticed that shade, so I just forgot about it, but it was very hot.

Understandably, staying in the shade or avoiding going into the river with wounded legs were both visceral. The team also reminded each other from time to time to drink enough water and apply sunscreen to prevent heatstroke and sunburn. Fairly speaking, even if Felicia had worked in the shade, the house would still have been built in the end. Interestingly, moving into the shade was seen as less appropriate, not in this weather, but in this social context where the volunteers were expected to be tough, to do what the team was doing, and thus practices like this were not considered too reasonable, or they avoided being judged as deviant from doing so. So, no matter how hot it was, Felicia decided to stay in the sun like everyone else. Despite the fact that she wanted to take up more challenges than she used to do in her daily life, staying in the sun in this weather was not her intention or expectation to achieve this. However, as the group was divided into small teams, they had their particular spots to do the task, which came together to build the house. This means that they had to fulfil the obligation at their designated spot. In this setting, it was less about self-interest or selflessness, it was more about actively doing belonging in order to be recognised as part of the group.

Moral community of like-minded people

Many of the informants were regular overseas volunteers or team leaders of different organisations. While some of them had already known each other before the trip, their previous experiences in volunteering gave them more topics to interact. Chatting among the volunteers connected people of similar interests and reconnected people in the same circle. Through these social interactions, some of them discovered that they had common friends or they joined service projects of the same organisations at different times. Dots were linked up into lines and then webs of social relationships, searching for status of being-with and being-in and expanding these moral circles of people with shared interests.

Before the projects started, the volunteers shared with others such as family, friends and colleagues, but most of them kept a very low profile except those who posted on social media asking people to contribute items to bring to Cambodia (for the ELIV team). They did not share with many people partly because the volunteers thought that other people did not have a clear idea about international volunteering. People were less interested in the volunteering itself, but more concerned about safety and health issues arising from the stereotypes about Cambodia as an under-developed country. Many volunteers felt frustrated when most people only imagined volunteering in Cambodia was doing something 'dangerous and dirtying their hands'. For example, Ilsa shared how her family reacted:

> when I told them, their response was like 'ha, going to that kind of place'... My family had zero concept, probably only knew about the Khmer Rouge, so they referred it to 'that kind of place', I guess they would have the same response for Cambodia, Vietnam and India.

She felt frustrated because the act of doing good was being muddled. After the trip, it became more meaningful for the volunteers as it was mostly about their experiences and achievements. However, sometimes they only received very cliché responses, for instance, Ilsa said 'they mainly sighed, "how unfortunate". When I said I would go again, they just replied "it is nice"'. The response was not so keen that the volunteers could not feel the recognition for what they had done. They felt more enthusiastic or even rewarding from sharing their experience with 'those who understand'. 'Those who understand' refer to people with previous experiences and similar attitude of international volunteering. They found it easier and more rewarding to talk to friends or colleagues with previous experiences in international volunteering as they are already in the same circle of interests, the wider moral community of volunteers. Since they have shared interests, or 'in the same field', they could talk in more details and feel for each other. These people are the appropriate listeners who would appreciate and recognise what the volunteers have done. Unlike other people who the volunteers found cynical about international volunteering, these listeners helped to justify the appropriate choice for developing the moral self.

'Those who understand' are also people showing interest in participation but encountering constraints in doing so. As the informants said, some of their friends wanted to join but were restricted by their job or financial conditions, therefore sharing with them was 'pleasant', because these people were willing to listen to volunteers' stories and respectful for their choice, and they asked for details of their experience. Their listening engaged both the storyteller and listener in a temporal space in which the storyteller (volunteer) fulfilled the need to tell others about the heightened experience, found the recognition because the listener responded positively, as well as reaffirmed what was believed by the storyteller. In this sense, they were actively and consciously creating belongingness or constructing this moral community through the recognition so as to further construct the moral agency and dispositions to volunteering as the appropriate choice. Amelie shared,

> for those who understand, such as Zoe. She came to me a few months before the trip telling me that she's interested, so I shared my experience with her... what we did, why I think it is great... Some may not spend money for volunteering, but they respect what we do. For example, one of my very good friends, I share with her every time after the trip about what we did there. I did ask her to join, but her financial condition is not so good that it will be a burden for her. But when I shared with her, she's very willing to listen.

Their listening means that their experience was being heard and their values were being agreed with. This somehow resonates with the act of moral selving to seek self-recognition. This act of sharing the stories is less about virtue signalling to reinforce their elite status or higher level of awareness as what Boluk (2011) found. They sought self-recognition because they had received more cynical responses or comments on their participation before the trip, partly a clash between old and new ethics. This might also be a way to recognise the qualities or values added to the body. The volunteers were expecting their listeners to feel the same excitement and probe into more details, and the right details.

Volunteers were also concerned about what and how to tell about their experience. Some volunteers wrote a lot of postcards to their friends telling them what they had seen in Cambodia, while some shared their stories through giving souvenirs. These postcards and souvenirs, originated in the place of their heightened experience, came with the volunteers 'back to the ordinary realm almost as a living messenger of the extraordinary' (Gordon, 1986, p. 140). The things they brought from Cambodia embodied their encounter with the people, and the '*tu di* 土地' (translated as *land or place*) as they termed, as well as the qualities accumulated from their extraordinary experience of volunteering. This was not only about the object but also the story or memory behind, therefore they wanted those who received the postcards or souvenirs to treat them extraordinary, or 'at least not ordinary' as Zoe said. For instance, when Gwendolyn gave palm sugar as a souvenir to her colleagues and friends, she included a handwritten card telling the story behind the palm sugar. On the card, she explained why palm trees were cut down in exchange for money and how the villagers forgot the importance of these so-called national trees, and what kinds of work was done by ELIV to assist the sale of the palm sugar to volunteers. She wanted to share about her trip through this physical form of embodiment while she also wanted her colleagues and friends to feel her emotions from the experience. She said,

> after I explained, their response made me feel a bit sad, I expected them to at least feel a bit amazed like saying 'really?!', but they were very calm. Later, I reminded them to keep the sugar in the fridge, and they just cared about how to handle it. They were not interested in the story, but I insisted in telling them.

Probably the content of the palm sugar was the same as those sold in supermarkets or souvenir shops, but the palm sugar wrapped in leaves from those palm trees of the village embodied her contact with the country although she never met the villagers nor that village.[1] Her insistence on sharing implied that she was seeking a sense of recognition and fulfilment from others. Some volunteers found it meaningless to share their experience with people who did not appreciate or support the idea of international volunteering, or who criticised 'why paying to volunteer'. This kind of criticism also mostly came from the older generation, as the volunteers found those of the same generation tend to be able to understand international volunteering. Amelie, for example, received such kind of doubt since her first service trip. Many times, she had been in the same situation in which she avoided explaining to the senior relatives or confronting them to defend her participation. While the volunteers like Amelie could not feel respected for their choice outside the moral community, they would rather actively strengthen her belonging to the moral community through continued participation.

Beyond the moral community of like-minded people, the volunteers did not find a spontaneous moment to start the topic and share their experience with a group who were not on the same trip, as it sounded too deliberate as if they were showing off. This echoes with a narrative discussed in other studies (e.g. Couch & Georgeou, 2017; Henry, 2019; Ong et al., 2018) that many volunteers easily fall back to the old everyday rhythms and life politics upon returning home. Although the examples here are not as extreme as that in the study conducted by Couch and Georgeou (2017) in which a respondent said 'I really felt that no-one understood me or gave a shit' (p. 28), there is a gap in expectations within and outside the moral community. Developing and expanding the moral community of like-minded people somehow helped them to bridge back to the everyday rhythms with someone listening to their stories, which they believed helped to shape the consciousness of issues that people in the moral community are concerned about.

Moral community with the locals

Although this study was looking at the practices of volunteers, they did not exist in the context alone, neither did they want to visit without having any contacts with local people. When asked about the differences between making a donation and volunteering, the volunteers stressed the importance of being-alongside and being-with the local people. Through the encounters, a moral community with locals also emerged.

When the volunteers were asked about the difference between making a donation and a service trip, many of them emphasised that making a donation only embodies 'doing something good'. One of the volunteers Dawn even classified different forms of doing something good, saying making donation can be the form when people could not go on a service trip for various reasons. Thus, making donation means the work of doing something good is done by the organisations, but then it raises the question of how the good is done. As criticised in the literature and by some sending organisations, sometimes things were not given or brought in for the right context or right people. For example, ELIV Project Cambodia team had seen many abandoned half-done toilets built by other NGOs in Svay Chek Village during their initial phase. It was because those NGOs expected the villagers to complete the work by themselves but the villagers did not have the resources or they did not know why and how to use and manage the toilets. I also observed that the bicycles donated for primary school children in the villages were too big for them, so they were riding in strange positions, which was not good for their growth. When Wendy and Ivan were discussing about the facilities in the school, they commented,

> the roofs were made of metal sheet, must be very noisy during rainy days. There were obviously no thoughts or planning, or they even had not been to Cambodia before. Otherwise, they should have known about the weather in Cambodia, or have talked to the principal so that they would have known what materials should not be used… if they really have this intention, this would have been avoided. But obviously, it's a product of donation from overseas.

Undeniably, the act of making a donation is not bad, but the volunteers believed that being there is important to have the good done more properly. Being present in the place enabled seeing what the local people tell them, how the local people live their lives, and what could be done. This seeing and being-alongside the local people helped the translation of need into authentic and relevant forms of care, like York said,

> to see what those people need, when we provide what they really need, appropriate, then the value is realised, but not giving them like a boat when there is no sea. The impact is the greatest when things are used in the places where they are needed the most, that's not a matter of the amount of money.

Donation can be made by sending a cheque or a click online, and this being-there could be achieved by the organisations' staff. It is not a matter of the amount of money but what that amount of money can bring to both the local communities and the volunteers. The presence of volunteers to participate in the means made the care in a more visible and tangible form as the outcome of the activities such as the house and toilet built, 960 mangrove seeds collected and potted, and 1000 mangrove seedlings planted into the sea, and reciprocated through the smiles from and hand-shaking with local people as a thank-you. This made the abstract form of care become more direct as it was done and felt by the volunteers contributing to their active self-formation. For the volunteers, moral fulfilment was embodied through tangible objects because the feelings could not be transmitted simply by sending a cheque or making an online donation. While donations could provide physical forms of care, they felt that warmth and personal connection are lacking to make those objects truly meaningful. For example, Hestia received a parcel on the day before our interview. She thought it was her regular online shopping, but then realised it was a gift for her support of an LGBTQ campaign, which she had completely forgotten about when she did it online. Erica also shared her thoughts on this,

> I think giving them money directly is not so good, so maybe in another form. It [the house] is the outcome of both of us putting into efforts,

> instead of doing nothing, just getting… giving money. They also contribute to some physical labour, work together, in exchange for a house, which is better than hiring people to do and it's like a gift from nowhere.

The house the family received was not simply a house. It was a house built together with the volunteers. Here, it is not about the absolute relationship between problem and solution, which is sending money and having the shelter problem solved. It is rather about the relational connection, which is the collective action in accomplishing the task of building a house and knowing what is being done is useful for the family. As the volunteers believed, also shared by the Project Manager, the people living in very remote villages may not be able to receive much information from the outside world, their social circle is limited to their village. At the same time, they found social ties are getting much looser in their daily life, so having opportunities to interact with these people in a Cambodian village is extraordinary. Their visits help to connect people back to the web of relationships, which cannot be achieved by sending donation. The volunteers perceived a community of themselves and local people embodying the fulfilment of collectivity that the volunteers were seeking for.

This morality of being-alongside was also conveyed by the organisations' staff. On various occasions, the local staff of the NGOs reified that the value of the action was not just the objects that the local people would receive at the end; it was the process of volunteers and local people working together to achieve the common goal, the local participation and the interaction in the time spent together that made the action meaningful. Zoe remembered Kiri, the translator of ELIV, once said,

> if workers are hired to build the house, all done by themselves, the story of this house is just about Cambodians, but now it's built by volunteers, the driver, locals, workers, all together… its meaning is more significant.

Jenny, the local Project Manager of ELIV once told us the feedback she got from the villagers, 'they [the villagers who receive the house or share the toilet] think it is a group of people coming over to help to do it together'. It is not simply a give-and-take relationship; it is a form of sharing which ELIV has emphasised through its projects. On ELIV's website, a quote from Jenny explains this idea of sharing:

> We also hope that through ELIV's programme, you will no longer define volunteering as unilateral assistance, but rather 'sharing' to respect each and every life, so that the story can become the driving force of 'life influencing life'.
>
> (ELIV, 2018, translated)

This corresponds to what Kiri said that the story is developed from collectivity, and 'life influencing life' is how such community is formed and one of the most appropriate ways of treating each other for changes, according to ELIV's philosophy of impacts. Without the visit of the volunteers, it would remain an I–It relationship (Buber, 1958) in which the people getting the house and the toilet experienced a detached thing, fixed in the location where they were built. On the other hand, the presence of volunteers and the interactions and collaboration created the I–Thou relationship (Buber, 1958) which was a more dynamic, living process of participating in the encounters with others, both by the volunteers and the local people. Although they did not have frequent and direct communication due to language barrier, the time they spent learning how to build the walls of the house by splitting and polishing the bamboos for the frame and then weaving the bamboo leaves into panels was part of the experience as all these parts were then assembled into the house.

This being-alongside, I–Thou relationship did not necessarily belittle the status of the local people if they are engaged in a rather harmonious relationship with a common goal and trust. Construction work of both service projects was led by a chief construction worker and assisted by other villagers who had the know-hows, here I used the ELIV example to illustrate. As it was the first time for most of the team doing any construction work, the volunteers had to follow the demonstration and instructions given by the locals and worked in small task teams. When the locals were taking the lead, rather than asking through the interpreters, the volunteers learnt how to speak in Khmer and then talked to the local workers directly such as when they wanted the local workers to demonstrate driving a nail into the bamboo frame. On the day the toilet was built, Bosco was part of the team that built the foundation with bricks and the whole structure with bamboo and metal sheets, so he worked closely with the lead worker, who we called '*Me Cheang*'. After the shadowing, he and *Me Cheang* quickly developed some sort of rapport despite their language barrier. At the end, *Me Cheang* appraised Bosco for picking up the skill very quickly, while Bosco thought he was working like an apprentice to him.

The whole process was led by the locals because of their skills and assisted by the volunteers. Admittedly, the volunteers were not there to make things perfect, but the outcome embodied the moral coevality and exchange. It has also created and produced stronger sense of community between volunteers and local people, which is conducive to bring them to the state of betwixt and between.

The communitas

Instrumentalising the collectivity

Oliver-Smith (2005) states that communities do not construct themselves but evolve. The moral community of volunteers was started by the sending organisations at the screening process and built upon by the volunteers themselves. Arrangements by the sending organisations helped this moral fulfilment through instrumentalising a sense of collectivity.

Kitchen was a significant setting fostering the intensity of being-alongside and being-in among the Hong Kong group. For the TBW Team-A, social interactions that happened in the kitchen were conducive to strengthening the sense of community. Coincidentally, each sub-team for the household tasks (cooking, washing up and cleaning public bathrooms) consisted of one volunteer who was good at cooking and three others who were inexperienced. As most of them wanted to learn, the time spent in the kitchen was more about having a cooking class than accomplishing an assigned task. Those from other sub-teams who were not on duty also gathered in front of the stoves for the cooking class, or just came to help setting up the tables for the meals. Preparing ingredients and washing up dishes were also intimate times when four or five people were sitting around, cutting and washing. That was the time when the team shared and discussed more personal issues, sought advice, or just made fun of one another. In the dining area, interestingly, after sitting on the floor for dinner in sub-teams for several days, the whole project team shifted to sit together at the large dining table.

Because of these settings, more social interactions that were not organised or instrumentalised happened beyond voluntary work, fostering this community to develop beyond the sending organisations' intervention. The spatiality of relationships was important in the sense that

> moral capacity is fostered, cultivated, and exercised within the social environment of the small-scale setting – or it is not acquired at all … what is fostered here, in the setting of proximity, is the capacity for developing empathy with others.
>
> (Vetlesen, 1993, p. 382; as cited in Smith, 1999)

In these micro-spaces, distance was reduced, both physical distance by proximity and moral distance by empathy. At the same time, moral codes were suspended in this temporal space, allowing the volunteers to act ordinarily, such as sitting on the floor full of vegetables and ants or bare foot being splashed with soapy water, and to open themselves up. Although the setting was constructed by the sending organisation, the volunteers built on and up the importance of this space. Morality was then grounded through and in practices of collective responsibility, so more intimate interactions and more 'we' relations with increasing intensity were formed.

The sending organisations also helped to engineer contexts for speeding up the team building process in order to enhance the intensity of experience which in turn facilitated the group work. It was not only the social setting that was important, but also the place and the people. For example, team building began in the shared bedrooms of TBW Team-A, where late-night conversations fostered familiarity and intimacy among small groups of volunteers. This enhanced rapport carried over to group interactions, promoting shared responsibility and creating a more harmonious and cohesive working environment. However, Ethan, one of the project coordinators, was not satisfied with the rhythm of team building as he saw gaps in the group cohesion, so he created the space to speed up the team building. He said,

> probably because we were divided into several rooms, we couldn't gather all the time like how we did last year, the team building process was much slower, it's almost until the 4th or 5th day … I gathered everybody in my room in the evening, people slept in the same room. The next morning someone said she got kicked, and another one said she got allergy. Afterwards, it's much more improved, Adam didn't get too much verbal attack, we pushed him a bit more, chatting more with him, so the team building was getting better. It's better to hang out in the same room, even if we didn't talk, it's still good to gather together and things would happen naturally.

As he said, even if they did not talk, staying in the same room already created a social setting of coevality. Everyone felt comfortable with one another's presence in the same room, and the web of relationships became more complex but solid. This temporary intimate space allowed them to experience everything together, even the silence, the idle time in the bedroom. Through this engineering of team building, people were able to find a conformable mode of being and to open up. Therefore, creating ways of being-with or being-alongside helped to fill up the gaps and thus the community was further constructed with growing closeness. This somehow enhanced the understanding of others which facilitated to produce a stronger sense of being-with, or communitas.

In contrast, the Taiwanese group lacked similar settings for the team building. It was partly because the activities took place in different locations of the city. It was also partly due to the group registration offer some people joined with friends. The established friendship had formed some small circles which posed some challenges to the group cohesion. For example, a Taiwanese volunteer said a few participants were always staying together, 'they had their own world'. In addition, I also observed that two other volunteers often stayed together. These small units did not interact intensively as a whole, creating resistance to the cohesion. Hestia, who has had very long international volunteering and leadership experiences, shared her feeling on this:

> I think this team, we are very willing to share, but I have been thinking what was missing. We missed something that connects us even after returning to Taiwan. A few of us will chat individually, but not the whole group… maybe six of us, but it's hard to find two extras to join as myself, Amelie, Gwendolyn and Zoe have known each other before the trip and we meet regularly… It's like three small circles within the team… I think group dynamics helps to pull people closer, and that's what I have been looking for from each volunteering trip. This time, it's fine as I have made many great friends, but it's a pity that we did not form a very close team.

Given the setup, the Taiwanese group 'improvised' in order to fulfil their quest for the collectivity that was lacking in their daily life. Commuting was one of the best times for this to happen. It took one hour to travel from the accommodation to the village and at least half an hour to other activity sites. Most volunteers consciously and proactively changed their seats every time to talk to different people on the coach. Still, the volunteers found the group very weakly connected, partly because of the diversity of the group. They came from diverse backgrounds, half were university students, a few were in postgraduate studies (including myself), some were working and one was about to retire. Many believed that since they did not have enough shared interests, the lack of common topics divided the group resulting in some becoming closer friends while the others being only 'part of the same team'. When talking about this, Felicia felt a bit embarrassed, as she did not feel integrated, however she still said, 'they are very nice'. Also, this socio-demographic diversity implies some sort of moral distance, or a sense of seniority in Asian culture, that the younger team members were hesitant to make fun of the elder ones, or at least their social interactions had to be proper, resulting in a loose sense of being-in. On the other hand, Vera, who has been an experienced team leader of ELIV, saw it as the *yuan fen* 緣份 (usually translated as fate, destiny, fateful coincidence, or predestined affinity). She said, 'if you got good *yuan fen*, then you would meet a group of nice people and get along well, otherwise you just try to maintain it during the trip'. Even though they could not develop it into a friendship, they kept themselves in the moral relationship and did what they should do as part of the group, as all these together still constituted their experiences.

This *yuan fen* also facilitated the development of the relationship among the Hong Kong group before the trip started. In pre-trip briefing and training workshops, the volunteers started working together to prepare for the activities of the trip. Discussions, either in person or online, provided many opportunities for the volunteers to get to know one another, socialise, and find the suitable ways of interacting with different companions. Many of the volunteers from the Hong Kong group said that they did not understand why but they felt they had known for long time when they met again at the airport on the day of departure. Rachel, who attended most of the pre-trip workshops, elaborated on this:

> In the first three pre-trip workshops, we were not too embarrassed or shy to talk to each other, we did talk. I'm sure we would not be close when we first met, but at the airport, we were like old friends who hadn't met for long time, maybe we knew that we would be spending the coming 12 days together, so naturally we did not have the awkwardness. It's like skipping part of the process.

This familiarity which was produced during the pre-trip preparation work possibly had helped created the sense of belonging, in the community formed by the sending organisations before the trip, before the actual work. They did not skip the process but because it was passive, intangible and ineffable (Bennett, 2012).

The time when they arrived at the airport signified the beginning of the trip, the actual volunteering experience, leading them to develop a sense of community within such an ambience. The boarding gate marked the literal entry into the liminality. In this liminal space, everyday practices, norms, obligations, roles and expectations were altered and suspended. Here, volunteers found it extraordinary as they were able to do what they should not or could not do in their everyday life, and they could care and be cared for in an inclusive and supportive space. As the informants pointed out, in-group relationship was developed into friendship and a sense of community was created quickly, partly because people in the volunteering space were willing to let go of their pride or persona in their everyday life. The communitas produced helped them to liberate their emotions and the authentic self in a community of trust and intimacy, at the same time created collective joy and unforgettable memories among the group in this extraordinary space.

Revealing the self in the liminal space

Volunteering offers an inclusive, liminal space that allows volunteers to liberate emotions often suppressed in daily life, fostering a search for the authentic self. In this leisure space, 'conventional social norms and regulations are often temporarily suspended as tourists take advantage of the relative anonymity and freedom from community scrutiny' (Kim & Jamal, 2007, p. 184). Free from everyday expectations – such as being a friendly colleague or a dutiful family member – volunteers can express emotions and thoughts without disguise. Although volunteers in this research did not encounter difficult existential questions as suggested in previous studies (e.g. Brown, 2013; Kontogeorgopoulos, 2017), this liberation itself is a form of self-care, not necessarily self-interested but essential for individual well-being. It is also the practice through which the individuals are conscious of their status in the formation of ethical subject of their conduct (Dilts, 2011). In this environment, freed from competition and conflict, individuals show more care and are more open to be cared. For example, Amelie, revealed that she was often preoccupied with others' opinions and crafted an image of a 'good' daughter at home and a 'good' teacher at work. During her volunteering experiences, she could share this openly, reflecting on her desire to uncover a self she truly appreciates. This temporary suspension of everyday identities provides volunteers with a unique opportunity to explore and express their true selves.

Hestia has also been exploring different dimensions of herself on service trips, and experimenting the one that suits different communities she belongs to. This was, as she explained, on the one hand to understand and unveil the better self, and on the other hand to enjoy the designated time-space without judgements and conflicts. She said,

> in my workplace we have much more office politics, whenever I see that group of people in the office, I don't feel comfortable… my state of

> mind… is totally different from that when I am working, as I have a lot of baggage in workplace, but it's very relaxing in Cambodia. That's the true me… I think it's because of the teammates… my state in Qinghai is much closer to the true self than in Cambodia. I should put it this way, every state is 'me', but the 'me' in Qinghai is the one I like the most, I am also 'me' in school, but I find the self that I like from volunteering, so it depends on the companions. If I can talk freely, that's because of the people around me. Amelie said she cares about how others perceive her, so do I. When people around me can accept this state of 'me', then I will reveal it; if I show the most comfortable 'me' but people are negative towards it, then I won't be that 'me' anymore in front of them. It's like to try out which state of 'me' suits that context. The state of 'me' in Cambodia is the second one I like the most.

During the time of volunteering, the moral codes of encountering difficult people in places full of competition, politics and conflicts were suspended, the volunteers were free from the bounded self in front of those people. Although they needed to suspend their own self to engage in collectivity and become part of the community, it was not a self-denial (Buber, 1958; Crossley, 1996); they had the safe space to seek the authentic self or experience different facets of the self while the normal presentation of the self was temporarily suspended from the quotidian (Goffman, 1959; Kontogeorgopoulos, 2017). This state of being-with stresses the relations with others and implies understanding of others, and reciprocity. Thus, being-with was enhanced by their being-in experience, as the belonging created facilitates the expression of the true self.

When people were free from much of their everyday pressure, daily rhythms of sleeping and waking up, eating and defecating as well as injury and ache became their main concerns about and among themselves. For instance, some of the Hong Kong volunteers were worried about constipation problem from the very beginning of the trip, and all of us had to 'check in' for our first stool until the fifth day when Gloria finally had hers. Everyone was very worried about her and at the same time teasing at her that she had eaten a lot from the first day. These physiological needs are so basic and visceral that they are never common topics for interactions in the daily life. These interactions in this setting were not simply expressions of biological necessity but were intimate experience with bodily rhythms of actions and genuine conversations with joy, which constituted part of ways in which the volunteers made a space for caring for themselves and others and at the same time a space for being cared (Milligan & Wiles, 2010). The collective joy produced contributed to the development of bonds and friendship among volunteers and nurture of I–Thou relations, which many volunteers felt were less likely to occur outside the liminal setting of volunteer tourism. This, in turn, produced heightened experience and a sense of belonging. For those teasing at themselves or revealing the true self, these were also active moves of doing belonging. In this space of caring and being cared, they did not need to hide the fact of suffering from constipation, or to feel shamed or embarrassed to share in public, which would otherwise be considered improper to talk publicly in everyday life. These seemingly minor issues embodied their freedom and fulfilment of their need to be heard and cared for.

Given a free and caring environment among the group, volunteers were not only more relaxed in showing the care-free side of the self, or not behaving properly, they were also more willing to share and face their past. There were moments when the team members revealed very personal stories and emotions to people they just met in this moral community. Among the Taiwanese group, one significantly intense, collective memory was when Camellia shared her past in the last reflection session of the trip. Each of us had to pick up one pictorial card about our past and one about our future, and then took turns to share. When it came to Camellia, she revealed her past based on the card 'Falling'. She was a key player in a cheerleading team at school, but she had to stop after she injured her spine in an accident during a performance. Since then, she had suffered from severe depression and attempted suicide, and had undergone a long period of medication and treatment. It was the first time after the accident that she opened up herself and revealed all her emotions that had been contained for a long period of time in front of others. As she spoke, she was crying, and all of us were hanging on to what she said, crying as well. Personal stories shared by everyone throughout the trip had emotionally connected one another, giving her confidence to speak up as she felt being cared for. She said,

> over time, we would get used to wearing a mask when interacting with others, you would treat others very well or politely, or maintain your image, and hold your real feelings. But this team created a very nice atmosphere that made me feel that I don't need to disguise myself or wear a mask. So, in those eight days, I felt like I had not spoken for a long time until then and wanted to say whatever I could, a kind of self-liberation and it felt great, and made me feel that these are people I can talk to about my personal things.

This was also the value of a community that the trust and reciprocated care produced the right atmosphere for them to liberate themselves, their emotions, and that revealing scars from their past was not bad or wrong. The value of this collective space was also the cultivation of moral capacity and the development of empathy with others (Vetlesen, 1993; as cited in Smith, 1999). The team was so intensely engaged that the bonding has been strengthened. Everyone was moved by Camellia's story. This was partly because they shared the same cultural background in Asian societies such as Taiwan where people are usually reluctant to express their feelings and emotions. This is also a norm in Chinese or even some other Asian cultures that people should not show their weaknesses. People tend to reveal and celebrate their success and veil their failure, especially to strangers. Therefore, this suspension of moral codes and release of emotions drew them closely together. York said,

> the most remarkable memory was the sharing by Camellia … maybe she had suppressed it for long, so sharing it with others is good for her. Sometimes,

people close themselves off, protect themselves, because of self-esteem, so as not to reveal the worst of themselves. Maybe it's okay to share in front of friends, but it's hard to reveal your saddest, ugliest or worst experiences in front of strangers.

In traditional Chinese families, communications between parents and children are usually not direct and explicit. As discussed in the previous chapter, parents tend to transmit the sense of responsibility implicitly rather than discursively. The traditional familial setting does not provide a space of openness. Thus, the volunteers were taking advantage of this freedom and openness in a volunteering space to express their emotions when others would not be judgemental. For example, Xena shared:

> I don't feel the same atmosphere in my family. Usually we follow what parents say, so we are very well-disciplined kids. My parents are very protective, sometimes we try to let them know that we have grown up, but the key is that we do not communicate so often that we seldom express our thoughts, it partially affects how I behave in school… It's somehow influenced by my family background as I don't have such practice at home. I want to practise how to express my thoughts or feelings, because it's not common in my family.

Being obedient to parents is a good quality of a child. However. It is believed by the younger volunteers that it is an old ethics. This liminal space provides a free space for them to listen to themselves, to be heard, to do something for themselves as a care for the self. It is also believed that nowadays the society has become more individualistic that people are more alienated, discouraging them to share their feelings and emotions and thus questing for caring in a reciprocal relationship between the cared for and the carer. This is vividly experienced by Camellia during her time of undergoing and overcoming distress and depression. As she said,

> not many people in my social circle in Tainan know about this [her accident and depression], even though a few knew about it, they just got to know but not really cared for me… In our education in Taiwan, you are never taught how to care for or to console others, so you will not know how to deal with it when you encounter such situation, people just know about it but will not say anything, they care more about their own. This team makes me feel that it is not a problem to speak it out. Such caring is very invaluable and hard to find in Taiwan.

The younger generation seems to experience a cultural confusion when they are under the influence of both traditional Chinese and Western values. On one hand, people tend to be implicit in caring about and for others although they are willing to show care, based on traditional values. On the other hand, people are getting more individualistic and responsible mainly for themselves. There has been a gap for this generation in the way of being in a social relationship, the way of expressing care, and the awareness that they can be cared by others. Thus, through these open dialogues in front of the group, what was fostered here was their moral capacity, their ability to develop empathy for others in proximity. This helps to fill the moral vacuum, the lost moral fabrics perceived by the volunteers. The ambience also created a safe space for the volunteers to share, to reveal their true self, to explore and seek their self-identity which might have been hidden or suppressed in their daily life when people are self-protected or care about their personal image.

Through liberating emotions and the authentic self, the volunteers entered the state of being with others through which they found their own being (Heidegger, 1949; as cited in Shubin, 2011). Sometimes, prejudice or stereotypes make people judge a book by its cover. The social interactions in such inclusive context perceived by the volunteers allowed them to reflect on their different ways of being in their daily life. As Camellia said, 'I got another stimulation that I cannot have such stereotype or prejudice'. It shows how volunteers returned home with a different perspective, bridging the ethics of care from the liminal space back to the ordinary time-space framed with their ordinary ethics. Their experience of communitas was so intense that it may feedback to their aspect of thought as well as dimension of action. This effect can possibly carry onto the aggregation stage of the communitas when they are returning with a new status or even actions for changes.

The relationship between volunteers and locals was shifting in and out of liminality, and in-between communitas and community. Various types of communitas emerged, some were spontaneous or existential while some others were more normative being organised into the permanent social system (Turner, 1969). While the volunteer-local relationship has been criticised as exploitative or unequal in the literature, the volunteers believed that they had a fair and equal relationship, particularly when they managed to interact like friends with mutual feelings of trust devoid of prejudice or assumptions. For example, Dawn said,

> I felt like I was not a volunteer, but it's hard to explain… it's an interaction between people, regardless of where you are from or your identity, simply the interaction between this person and that person at that moment, a fair relationship.

This was the communitas developed in the liminal space, everyone was treated equally, regardless of their socio-cultural backgrounds. More specifically, this was the existential or spontaneous communitas between volunteers and locals, as it was more transient in nature especially during such kind of counter-culture happenings (Turner, 1969). Some informants said that they were more like one of those transients of villagers' life, unlike the relations with the interpreters and local Project Manager as they had more contact and interactions.

Therefore, when meeting the villagers, the volunteers were still maintaining some of the moral codes from their daily life. For instance, Zoe was careful in managing the representation of herself as an outsider in front of the villagers. Due to language barrier, she said, 'we used gesture, body language, smile, you have to smile, otherwise look grumpy, as they could not understand what we were talking about'. In these kinds of relations, the social structure was not so loose that cultural norms were in place; but the partial connection and attachment, or cooperation kept them being in a community. Despite that, there were moments of engagement of both volunteers and locals in collective tasks which trust was developed while some of their suspicions towards strangers were suspended in their casual interactions. When the TBW-Team A was doing the seedling nursery, some villagers gathered around and the two local volunteers were also working together, so communication was possible with their help with interpretation. The villagers were making fun of a few of us, feeding us snacks as our hands were full of mud, or giving advice by the side. The moments when they gathered around and interacted with the volunteers created a collective joy. Through those interactions, Bella said she could feel the trust between the volunteers and the locals,

> they were very willing to share. We had an interpreter with us, they shared a lot, I could feel they were very sincere. Some even treated us some snacks, showed us some photos, not wary at all, didn't see us as strangers… When we went to buy coffee from the tuck shop, the owner was very friendly, she always taught us Khmer. They really wanted to have some sort of interactions with us.

Volunteers were surprised that the locals were very willing to share with strangers, a lot about their personal stories, meanwhile the volunteers also told the locals much about themselves, which they normally would not do in Hong Kong or Taiwan. As the locals revealed about themselves, the trust and thus the emotional connection was so intense that they talked further and deeper, without disguising themselves with their persona in ordinary life or being bounded by their moral codes of not talking about personal shames or stigmas. Both the volunteers and the locals wanted to create 'we' relations out of the collective settings, although they would possibly be temporary, which however depends on the intensity of the relations and how the collective joy is to be strengthened when they have to be apart. One palpable example was the farewell dinner that TBW-Team B organised with the teachers. A very young part-time teacher of the kindergarten went to talk to different volunteers during their stay in the village. At the dinner, she shared her story that she had to work in another country for a higher salary to share the financial burden of the family. Everyone was crying when listening to her story, carried away by her, at that time they felt they got closer as they were sharing something really personal. When the two teams returned to Phnom Penh, Selina had tears in her eyes when she shared this experience with the group. She told me later:

> Before the trip, I did not expect a local would share everything about herself with us, she cried so badly, very sad, and she was also unwell. She felt so sad to leave her family. This time, I realised our conversation could go so deep… I think our interactions made them feel that we were friends, closer.

This is similar to Camellia's liberation of genuine emotions and the true self in front of strangers. Here, the local teacher did not feel inferior when compared with the volunteers coming from Hong Kong, neither did she feel ashamed to unveil her saddest story and deepest feelings to them. The communitas was produced from the accumulated trust from their interactions, and this communitas was not so spontaneous, it was getting more of normative communitas when their relationship was growing from the incidental volunteer-local contact to friendship. In order to maintain the connection or relationship, the team gave the girl a memento which she could bring with her to another country. Melanie shared,

> we wanted to give her an HKU hoodie but it's too dirty with concrete and paint on it. Quinn had several guitar picks, he gave his favourite one to her… now we still keep contact on Facebook, messaging to see what happens on each other's side.

Despite the physical distance, they wanted to maintain the experience of being-with and being-in the community over time and distance, or at least the guitar pick marked the trip, their intimate time, their time being with each other.

Communitas over space and time

Communitas is not vanishing right after the trip; it is not on our timescale but lingers in its own way as time passes, in and out at different spatiotemporal coordinates. While the intensity of sentiments may be reduced with increasing distance (Ginzburg, 1994), relationships can be maintained over greater distance with the support of the global sense of place (Davies & Herbert, 1993). The communitas produced helped to hold the dots in the web of relationships together, strengthening the bonding and creating more I–Thou relations. This results in experience with great intensity because the volunteers and locals fully revealed their emotions, their true self, and all the basics became the most important. Although the trips were very short, the bonding developed was far stronger than expected due to the intensity and active moves of doing belonging. Ethan returned as a volunteer and project coordinator because of the meaningful relationship developed. He said,

> …the 10 days were like one year there. And they are like oldies, that's weird thing about bonding developed through service trip, because 10 days were not just like 10 days, it's the time we spent together, we got to know one another, bit by bit the bonding was very deep… It's the second year of our relationship, we all still have the initiatives to organise gatherings, to maintain our relationship. Even the locals asked about the rest of the team [from last year], asked how the rest of the team are when a few of us visited them again this year. The bonding is real, and that made me join again, in the organising committee to promote this, as helping others to build a network like this is also something worthwhile and meaningful.

The previous experience of Ethan was so intense that he decided to join again in building and expanding this kind of network. This moral community has evolved in the way that fosters the continuation of participation in volunteering or working towards causes concerned by the volunteers (McGehee & Santos, 2005). Among the Hong Kong volunteers, five of them joined the service trip organised by TBW in January 2016, and they joined the same organisation again as project coordinators of the projects in 2017. After their first experience, they stayed in touch; working together and sharing about the second experience with the rest of the group also helped drawing the whole group together, as if they were going on the trip together again. The collective joy made these five volunteers join again, while the communitas folded over and back reminding their belonging to this deep relationship formed in the year before. Their previous experience created their relations with this country, *tu di* 土地, embodied through the house remained on it. This *land* has become somewhere they know, with people they met and interacted, in an inner concentric ring of ethics. Returning to this place by some of the people in the group retrieved all sorts of collective memories and the doing of good. Despite the physical distance, the communitas developed and maintained over time and space reinforced the emotional connection with the country they visited and the people they met together, sustaining the moral community in which the moral subjects continue to participate alongside.

In order to sustain the joy from going on the service trip and search the best self they would like to see, Amelie, Camellia and Hestia signed up as ELIV's team leaders. During their training, they were supporting each other, exchanging information and feedback on different sessions. They also shared in our Line (a messaging app) group about their training and later their first leading service trip. This has helped to consolidate the bonding created during our trip, both among themselves through more shared experiences and among the group through emotional connection. Wendy also attempted to maintain the relationship by encouraging some of the volunteers of her previous teams to apply for team leader position, as she believed that by staying in the same moral circle of like-minded people, their friendship would grow, accumulating social capital simultaneously. Their extended involvement also brings them into another circle of volunteers and industry practitioners, extending their social networks and keeping this moral community to evolve. Wendy, who has been volunteering and leading service trips since her undergraduate studies, said,

> a good relationship will further develop. I will ask some of them to be team leaders, so we can lead together, go to Cambodia together or do something, we will become closer. Some even become very good friends. I did not become good friends with other volunteers except those who have become team leaders, only then can we have more opportunities to get in touch and talk more deeply, have more common topics.

Good relationships would mean those that can sustain the sense of belonging, the process of moral selving in the community. The service trip is an extraordinary space and time for trust, intimacy and deep interactions to be developed. Gwendolyn gave a similar account of how her relationship with other volunteers has been maintained due to shared vision of moral actions. She said,

> we met up from time to time, we organise gatherings for different districts, such as reunion or sharing in Taichung, then we would talk and become closer. Through regular team leader training, we had some activities and sharing sessions, sharing about values and accomplishing something together, it's a kind of camaraderie, then we get to know each other better. After all, a few of us would find that we share a lot of interests or have more common topics, then we may have gatherings more often.

Gwendolyn and her companions have organised reunions to maintain the conversation on volunteering. Although they joined projects from different organisations in between, their connection has not been discontinued as they still actively maintain their sense of being-in this moral community, making them closer and strengthening their friendship. Their ongoing relationship has maintained their social capital accumulated through the network and drawn them to volunteer together again. The evolution of the moral community further develops over time. Being part of the community allows them to find and maintain relationships with the like-minded people, to engage in the means which leads to the better self as the outcome. They are further connected with more act of doing good at the aggregation stage, which leads to deeper, more meaningful relationships.

In comparison, returning volunteers or team leaders of the same organisation tend to develop a closer relationship with the local staff in general, making communitas grow into a stronger sense of being-in-the-community. Wendy started as a volunteer and later became a team leader of ELIV since 2011, over the time she has developed a closer relationship with the locals so that their conversation was more intimate and personal. She recalled one of her experiences,

> I always want to sit next to them [local people] during lunchtime, squatting. One time, the driver, interpreters, workers, villagers and team leaders, all of us were sitting together on the ground for lunch. After eating for a while, Me Cheang started to ask me, 'when are you getting married, I haven't seen your boyfriend even you have come so many times? Do you have a boyfriend?'

This hints a temporal dimension of belongingness. Although Wendy or other returning volunteers do not maintain frequent contact with the locals after each trip, they still have this emotional connection. The connection seems to be on hold when the volunteers return home, and it re-emerges when they volunteer through the same organisation. This is because such relationship does not exist in our timescale of ordinary life, the community is sustained through the materials such as the house, the toilet, the mangrove trees and the guitar pick which contains the memory and traces of collectivity and their sense of fulfilment. The encounter and experience in tourism space has elicited emotional responses (Gibson, 2009), which is maintained by the ethics of care and the 'we' relations created.

The locals did not see the volunteers as outsiders, the others, while the volunteers made themselves become part of the others and gained the sense of belonging, or constructed this community with the local people from the mode of being temporary in proximity to being sustained over distance. The collective joy they shared reminded them of the experience of doing good together, the relation with the country because of those people they worked together. The intense revelation of emotions and personal stories heightened the experience through showing the hidden persona to strangers. This helps to develop their moral self, which also shapes their identity and belonging developed from these moral communities. Further, some volunteers were actively making belonging to the country by emphasising their relation of their body and mind to the country and people, for example, Dawn said 'having been there makes me feel like I am having blood from that country…'

Identity can be perceived as a self-conception, a presentation of the self. It is defined as 'how we make sense of ourselves… It refers to lived experiences and all the subjective feelings associated with everyday consciousness' (Rose, 1995, p. 88). From how the volunteers talked about their experience, their relationship with the places they visited and people they met, they have developed a sense of belonging to that country, because of their experiences of 'being' and 'knowing'. During a meeting with the Project Manager of an NGO that supports TBW projects, he asked Ethan what made him continue to coordinate the project to Cambodia. Ethan said, 'I like the people there, I think a part of my identity comes from there'. The Project Manager shared the same feeling, as he gave up his well-paid managerial position in an auditing firm and returned to the NGO as the Project Manager because part of his identity was developed in Cambodia since his first visit that kept lingering to remind his passion. After their experience of presence in that country and spending time and working with locals, they felt things in that country were 'part of you'.

Moral communities and social relationships

Bellah (1995, p.50) states that

> [a] good community is one in which there is argument, even conflict, about the meaning of the shared values and goals, and certainly about how they will be actualized in everyday life. Community is not about silent consensus; it is a form of intelligent, reflective life, in which there is indeed consensus, but where the consensus can be challenged and changed—often gradually, sometimes radically—over time.
>
> (as cited in Arai & Pedlar, 2003, p. 192)

These moral communities have actualised this quality of a good community by creating a space in which differences and contradictions are simultaneously enacted, challenged, negotiated and remained. This was showed by the negotiated sense of morality through the examples of Adam and Quinn. This was also evident in how they differentiate themselves outside the moral communities. On the one hand, the volunteers built up and expanded the web of relationships by acknowledging membership of other like-minded people, people of the wider community with shared interests. On the other hand, they differentiated themselves from people whose moral values did not go in line with theirs as they did not receive the recognition as desired.

A community is a microcosm of wider society. The quest for collectivity and the experience of being in meaningful social relations implies a lack of it in their everyday life, which is also clearly voiced by the volunteers. It speaks to one of the communitarian critiques – alienation and loss of community – of the current fragmented society. 'Community' touches very deep human aspirations to experience a sense of belonging (Mulligan, Steele, Rickards, & Fünfgeld, 2016), while a moral community entails a 'good' relation with the self, others, and the place. Here, a community is not one-dimensional, static, place-based. The volunteers as the moral subjects being crafted are travelling across different moral worlds to seek their recognition and the lost ties and care between people in the neoliberal times. Social relationships in different moral communities are not just encounters of others but with others. They are more of dialogical than monological (Buber, 1958), as they are engaged in communication and response, a reciprocal relationship. It is dialogical as the moral subjects need to feel it and participate by being together with the others they encounter.

It is regarded as a good relation as the volunteers are able to seek self-identity, to discover different versions of the self, to care and to be cared through the interactions. Their identity reflects the crafted moral self throughout the process of good-doing in group endeavours. These group endeavours created moral communities that allowed the volunteers to liberate the various dimensions of the self and are able to open up, engaging in deep and meaningful relationship through intense experience. These moral communities also serve as the collective from which the volunteers find their existence, being the individuals in relation to others. They also make attempts and active moves to be accepted to these moral communities or feel belonged to. This explains why simply sending a cheque does not work, as the cheque itself does not embody the encounter, care, belonging and recognition. It is not an input-output equation. The outcomes that the local community get in physical forms could be pretty similar from the amount of money a volunteer pays for the service trip or sending a cheque of the same amount. But what makes the difference here is the social relationships cultivated through growing mangrove seedlings, building the house, and revealing the inner self together, and thus the moral self that the volunteers want to develop.

This I–Thou relationship reconnects people in the fragmented society, rebuilding the lost moral fabrics. This sense of fulfilment is not simply the positive outcomes on the volunteers as summarised in the literature such as adding something on their resume. When collectivity and searching for the self is the process and the moral community and communitas produced is the outcome, it probes us to reflect on some of the issues in the contemporary Asian societies like Hong Kong and Taiwan that social bonding among people is generally perceived to be loose and there is a need to reconnect people. The return to community is to recall the importance of collective endeavour, to confront some of the social crises, crisis of identity and crisis of alienation, and to reignite and restore civic engagement (Arai & Pedlar, 2003). Collective networks formed embody the reciprocal social relations that builds up the social capital (Siisiainen, 2003; Stebbins, 2001). The communitas and the moral relationship helps sustain the moral actions through their experience of collectivity, mutuality and care. The moral lack in the society leads to participation in such moral actions, during which the belonging to moral community, either active or passive, keeps the individuals in the state of 'being-in' leading to more participation. This re-instilling of moral values or ability to interact with people to form web of relationships through values of community such as integration, reciprocity and trust helps us to rethink the moral landscape, what old and new ethics means in the contemporary times.

While the community of volunteers is a form of post-place community in which solidarity and shared norms tie people together without a designated locale, place is indeed significant as the place makes the relationship meaningful. Place here is not the locality or home country in traditional notion of community; it is what the volunteers referred to the *land*. This place becomes important and meaningful after this moral community is formed as it is where all these sorts of physicality and doing belonging take place. It is a place in which some of their moral codes and everyday restrictions are suspended.

It requires continuous doing of belonging and moral relationship, embodied through things brought over, built, left and took, or very ordinary things such as sweat, poop, insects and rotten vegetables which have not been talked about extensively in the literature. All these constitute collective memories and intense experience of being-alongside, being-with and being-in, the material outcomes or traces of the intangible attachment remained on that *land*. In addition, the people who shared the group endeavour, the understanding of the self and others, and from whom the volunteers sought recognition, also played a key role in making the relationships meaningful and the experience intense. In such social context, together with place and people, moral fulfilment is actualised.

This doing of belonging engages people in social relations which enable and benefit moral selving through taking up moral responsibility and reciprocity. Moral selving does not necessarily emphasise the self. The moral selving process here speaks how the development of the self is achieved through reconnecting the lost moral fabrics in our fragmented society; it is a fulfilment of bettering the social relations and gradually eradicating the social crisis of alienation. In the liminal space of volunteering, the temporal rhythms of living in Cambodia such as walking in the river, weaving a wall of bamboo leaves, drinking coffee, and learning Khmer embody the escape from the everyday pressure and other social problems. The time being on this *land* creates the right rhythms, positively interrupting the ordinary rhythms. At the aggregation stage, whether this temporal embodiment will be interrupted by the competition and pressure, or sustain to interrupt the ordinary status, mentality and actions and things get re-set, all depends on what the moral relationships will bring to the volunteers and move beyond. This offers a useful lens to understand how the moral self is cultivated and experienced with relation to others. The new experience of the self may induce alternative ways of living in different patterns and forms of care, responsibility and obligation. This may have an influence on the life-course decisions, together with their past experiences and anticipation of the future (Bowlby, 2012; Turner, 2012), if the volunteering space is effective in bringing out critical reflection and plans of making changes. It is worth moving further to discuss whether and how volunteer tourism can support this transformative process.

Note

1 The palm sugar is produced by villagers of a neighbouring village. The story was shared by the ELIV local Project Manager during one sharing session.

References

Arai, S., & Pedlar, A. (2003). Moving beyond individualism in leisure theory: A critical analysis of concepts of community and social engagement. *Leisure Studies*, *22*(3), 185–202. doi:10.1080/026143603200075489

Bellah, R. N. (1995). Community properly understood: A defense of "Democratic Communitarianism". *Responsive Community*, *6*, 49–54.

Bennett, J. (2012). *Doing belonging: A sociological study of belonging in place as the outcome of social practices*. PhD thesis. University of Manchester. UK.

Boluk, K. A. (2011). Fair Trade Tourism South Africa: Consumer virtue or moral selving? *Journal of Ecotourism, 10*(3), 235–249. doi:10.1080/14724049.2011.617451

Bowlby, S. (2012). Recognising the time – space dimensions of care: Caringscapes and carescapes. *Environment and Planning A: Economy and Space, 44*(9), 2101–2118. doi:10.1068/a44492

Bowring, F. (1997). Communitarianism and morality: in search of the subject. *New Left Review, 1*(222), 93–114.

Brown, L. (2013). Tourism: A catalyst for existential authenticity. *Annals of Tourism Research, 40*, 176–190. doi:10.1016/j.annals.2012.08.004

Buber, M. (1958). *I and thou* (R. G. Smith, Trans.). Edinburgh: T&Clark Ltd.

Couch, J., & Georgeou, N. (2017). Do immersion tours have long-term transformative impacts on students?: A study of Australian university students in a Tibetan host community in India. *Journal of Applied Youth Studies, 2*(2), 18.

Crossley, N. (1996). *Intersubjectivity: The fabric of social becoming*. London: SAGE Publications.

Davies, W. K. D., & Herbert, D. T. (1993). *Communities within cities: An urban social geography*. London: Belhaven Press.

Dilts, A. (2011). From 'entrepreneur of the self'to 'care of the self': Neoliberal governmentality and Foucault's ethics. *Foucault Studies*, (12), 130–146. doi:10.22439/fs.v0i12.3338

ELIV. (2018). About us. Retrieved from https://eliv.org.tw/aboutus/

Etzioni, A. (1995). *The spirit of community: Rights, responsibilities and the communitarian agenda*. London: Fontana.

Gibson, C. (2009). Geographies of tourism: (un)ethical encounters. *Progress in Human Geography*. doi:10.1177/0309132509348688

Ginzburg, C. (1994). Killing a Chinese Mandarin: The moral implications of distance. *Critical Inquiry, 21*(1), 46–60.

Goffman, E. (1959). *The presentation of self in everyday life*. New York: Anchor.

Gordon, B. (1986). The souvenir: Messenger of the extraordinary. *The Journal of Popular Culture, 20*(3), 135–146. doi:10.1111/j.0022-3840.1986.2003_135.x

Heidegger, M. (1953). *Being and time* (J. Stambaugh, Trans.). Albany: State University of New York Press.

Henry, J. (2019). Directions for volunteer tourism and radical pedagogy. *Tourism Geographies, 21*(4), 561–564. 10.1080/14616688.2019.1567577

Kim, H., & Jamal, T. (2007). Touristic quest for existential authenticity. *Annals of Tourism Research, 34*(1), 181–201. doi:10.1016/j.annals.2006.07.009

Kontogeorgopoulos, N. (2017). Finding oneself while discovering others: An existential perspective on volunteer tourism in Thailand. *Annals of Tourism Research, 65*, 1–12. doi:10.1016/j.annals.2017.04.006

Latimer, J. (2013). Being alongside: Rethinking relations amongst different kinds. *Theory, Culture & Society, 30*(7–8), 77–104. doi:10.1177/0263276413500078

McGehee, N. G., & Santos, C. A. (2005). Social change, discourse and volunteer tourism. *Annals of Tourism Research, 32*(3), 760–779. doi:10.1016/j.annals.2004.12.002

Milligan, C., & Wiles, J. (2010). Landscapes of care. *Progress in Human Geography, 34*(6), 736–754. doi:10.1177/0309132510364556

Mulligan, M., Steele, W., Rickards, L., & Fünfgeld, H. (2016). Keywords in planning: What do we mean by 'community resilience'? *International Planning Studies, 21*(4), 348–361. doi:10.1080/13563475.2016.1155974

Neville, M. G. (2008). When poor is rich: Transformative power of I-Thou relationships in a Brazilian Favela. *Gestalt Review, 12*(3), 248–266.

Oliver-Smith, A. (2005). Communities after catastrophe: Reconstructing the material, reconstituting the social. In S. E. Hyland (Ed.), *Community building in the 21st century*. (pp. 45–70). Santa Fe: School of American Research Press.

Ong, F., King, B., Lockstone-Binney, L., & Junek, O. (2018). Going global, acting local: Volunteer tourists as prospective community builders. *Tourism Recreation Research*, *43*(2), 135–146.

Rose, G. (1995). Place and identity: A sense of place. In D. Massey & P. Jess (Eds.), *A place in the world? Places, cultures and globalisation* (pp. 88–132). Oxford: Oxford University Press.

Shubin, S. (2011). Travelling as being: Understanding mobility amongst scottish gypsy travellers. *Environment and Planning A: Economy and Space*, *43*(8), 1930–1947. doi:10.1068/a43613

Siisiainen, M. (2003). Two concepts of social capital: Bourdieu vs. Putnam. *International Journal of Contemporary Sociology*, *40*(2), 183–204.

Smith, D. M. (1999). Geography, community, and morality. *Environment and Planning A*, *31*(1), 19–35.

Stebbins, R. (2001). Serious leisure. *Society*, *38*(4), 53–57. doi:10.1007/s12115-001-1023-8

Taylor, C. (1991). Cross-purposes: The liberal–communitarian debate. In N. L. Rosenblum (Ed.), *Liberalism and the moral life* (pp. 159–182). Cambridge, MA: Harvard University Press.

Turner, E. (2012). *Communitas: The anthropology of collective joy*. New York: Palgrave.

Turner, V. (1969). *The ritual process: Structure and anti-structure*. London: Transaction Publishers.

4 Transformative learning

Potential of volunteer tourism for re-inventing the moral self

Transformative potential of volunteer tourism

Conceptually and methodologically, this study did not start out to look at transformation in particular; rather, it was set out to understand why people travel as a volunteer and what they are looking for. Being aware of 'transformation failure' identified by Zavitz and Butz (2011), this chapter takes a step back. It focuses on how the moral self may be developed from the activities and practices during volunteer tourism, both planned by the sending organisations and encountered by the volunteers in dynamic settings. This widens the analytical lens to the spatial and temporal antecedents and parameters and opens up the discussion on the transformative potential of volunteer tourism. In this chapter, I am going to examine the potential of volunteer tourism in transforming volunteers' frames of reference, and if postive, how and what kind of transformation takes place. This chapter is intertwined with Chapter 5, as they both focus on transformation. This chapter stresses more on the experiences using the transformative learning theory, while the next chapter offers a closer look at how the sending organisations play a crucial role in reshaping the orientations of the moral self.

In Chapter 1, transformative tourism has been reviewed with reference to the hopeful tourism as well as the conscious self-formation discourse. The concept of transformation has been primarily discussed in education using transformative learning as a framework for understanding adult learning. Mezirow's transformative learning theory (1991, 1996, 1997) explains how individuals experience shifts in their frames of reference – the underlying beliefs, values, and assumptions that shape perceptions, feelings and actions. These frames consist of habits of mind, broad habitual ways of thinking, and points of view, specific attitudes and interpretations based on personal and social influences. Transformative learning occurs when individuals become critically reflective of these frames, leading to meaningful changes in perspective. However, these frames of reference can often be problematic or taken for granted, shaped by fixed interpersonal relationships, cultural biases, political orientations, occupational habits, and other societal influences (Mezirow, 2003). These ingrained perspectives may limit an individual's ability to engage critically with the world around them.

DOI: 10.4324/9781003413929-4

The key to transformation lies in the ability to become critically reflective of these assumptions and to learn how to negotiate and act based on one's own purposes, values, and meanings, rather than uncritically assimilating those of others (Mezirow, 2000a). This transformation occurs when an individual confronts new information that disrupts their meaning perspectives, which are being shaped by epistemic, sociolinguistic, psychological, moral-ethical, philosophical and aesthetic factors (Mezirow, 1991, 1997). Such confrontations prompt individuals to construct and adopt new, revised interpretations of meaning based on their experiences.

This process of transformative learning can lead to a three-dimensional change, namely psychological shift (a deeper understanding of the self), convictional change (a shift in beliefs and ideologies) and behavioural transformation (modifications in behaviours and lifestyles) (Mezirow, 1996). There are 10 proposed steps in achieving transformation (Mezirow, 2000b, p. 22):

1 A disorienting dilemma
2 A self-examination with feelings of fear, anger, guilt or shame
3 A critical assessment of epistemic, sociocultural, or psychic assumptions
4 Recognition that one's discontent and the process of transformation are shared and that others have negotiated a similar change
5 Exploration of options for new roles, relationships, and actions
6 Planning of a course of action
7 Acquisition of knowledge and skills for implementing one's plans
8 Provisional trying of new roles
9 Building of competence and self-confidence in new roles and relationships
10 A reintegration into one's life on the basis of conditions dictated by one's perspective.

In contrast to Mezirow's phychocritical perspective, O'Sullivan (2002) defines transformative learning with a focus on a more sustainable lifestyle through 'experiencing a deep, structural shift in the basic premises of thought, feelings, and actions. It is a shift of consciousness that dramatically and irreversibly alters our way of being in the world' (p. 11). Taylor (2008) has also suggested three components of transformative learning that are important during the process, including self-reflection, engaging in dialogues with others, and intercultural experiences. One common component of these various conceptions of transformative learning is critical self-reflection which leads to changes in the frames of reference. This then highlights the importance of discourse which, as Mezirow (1997) explains, is 'a dialogue devoted to assessing reasons presented in support of competing interpretations, by critically examining evidence, arguments, and alternative points of view' for meaning-making (p. 6). Discourse with others in a different or new culture is the medium through which transformation is promoted and effected (Taylor, 2008), which would lead to a perspective transformation when the 'unfamiliar' experiences help to question the 'familiar' everyday situations and practices (Bamber & Hankin, 2011; Mezirow, 1997). Different types of learning are also needed for different types of goals. Instrumental learning helps to achieve short-term goals while communicative learning is useful for long-term objectives (Mezirow, 1997). This could help us understand the instrumentalisation of service obligation in schools in Hong Kong and Taiwan, and how the service-learning approach could be further developed into communicative learning to achieve the long-term goal.

Transformative learning theory has been applied in understanding specifically volunteer tourism among many other forms of alternative tourism (e.g. Coghlan & Gooch, 2011; Hammersley, 2014; Knollenberg, McGehee, Boley, & Clemmons, 2014). Coghlan and Gooch (2011) found that there is still a lack of existing volunteer tourism programmes that incorporate the two steps of 'planning of a course of action' and 'provisional trying of new roles', which they believe are useful in moving volunteer tourism beyond the rhetoric of 'doing something good' to more life-changing or transformational changes for participants and the wider community of people. Hammersley (2014) also extends the conversation by arguing the importance of structural opportunities for both pre-trip preparation and post-trip debriefing of volunteers. These processes facilitate the transformation, during which the expedition leaders would play a key role in providing support through the process of disorienting dilemma or effecting changes in the frames of reference (Taylor, 2008). These provide the basis to further examine whether and how transformation can be actualised in the case of volunteer tourism.

Experiencing dilemma

In the transformative learning theory, a disorienting dilemma (step 1) could be an extraordinary ethical dilemma, but also an ordinary situation such as taking a shower. In Cambodia, the volunteers fell into dilemmas from how they perceived local resources and their ways and feelings of using them. After the pre-trip sharing by previous volunteers who worked in the same village, the TBW volunteers said they were mentally prepared for having limited resources especially water for their daily use. The volunteers, including myself, heard how taking a shower was like the year before: they took shower in the toilet where insects were flying around at night and the water in the water drum was muddy. So they either did not take a shower every day or rinsed themselves (with clothes on) with a dipper outside the toilet. Ona of TWB Team-B, a returning volunteer to the same village, experienced a dilemma and feeling of guilt when she saw three big drums of clean water prepared exclusively for the volunteers. She and a few other volunteers had the same dilemma of whether they should wash their hair every day. The dilemma came from the fact that three drums of clean groundwater were very precious in dry January in Cambodia. Washing hair was so ordinary that they did it every single day in Hong Kong without questioning the need or the constraint of doing it. She said,

> I did question about it, like Cambodians do not wash their hair every day but Hong Kong people can't stand with it. We got 10 people, there were three drums of water outside, they kept refilling the drums for us with groundwater. I wondered if that would cause much trouble to them as they do not have enough clean water resources, I am not sure about that, but we used a lot of their water for showering. It seems we did not really value the water, splashing a huge scoop of water to wash our hands, not caring how much we used.

On the same issue, Selina, who was also a returning volunteer to the village, felt that they were using the clean water more responsibly by adjusting their practices this time. She said,

> when we arrived, we saw those three drums of water. You know, 10 people, based on how much water we usually use for a shower in Hong Kong, we would use a lot of their water, in addition to the amount used for cooking and washing up. I saw that everyone was aware of this issue, trying not to use too much or not to shower for too long, as long as the body got wet and cleaned, trying to use the minimum, especially the girls with long hair, we just did it quickly without using a lot of water.

Both Ona and Selina experienced a feeling of guilt but negotiated in different ways. Ona was undergoing a time of self-questioning based on the practical local situation; clean groundwater was limited in the dry season while the temporary stay of 10 people used a lot of the water. On the other hand, Selina looked at the relative conditions and practices. Of course, they could not shower or wash hair under running water from the shower head as how they did back home. The way she negotiated the guilt was to use less water by speeding up. Therefore, Selina thought that they were aware and adjusted accordingly in a relative sense that they used less water than their usual practice at home rather than the absolute sense that they decided not to shower because of the scarcity.

Similarly, TBW Team-A experienced the awakening point from taking a shower in the fishing village. There were big water tanks storing rainwater and groundwater for tourists. Having heard the same story at the pre-trip sharing, they had also expected the worst that they would have no water until they saw a good flow of water from the shower heads, otherwise they would just use a towel to wipe their bodies, as Kathleen planned to do. Halfway through their stay, they realised that the water tanks were almost empty when the shower started to drip. Again, some of them did it faster or splashed two scoops of water from the bucket just to feel like they had showered instead of not showering. When they returned to Phnom Penh and later to Hong Kong, they noticed the huge difference in the amount of water they could get from the shower, which they had taken it for granted in their everyday lives. Without these felt visceral experiences, they would not have known or even questioned whether they had used more of the water than necessary, and other issues such as the resources in other places and their practices in an unfamiliar context. The ordinary shower became extraordinary when they had observed the amount of water used differed. Ona, Selina and Kathleen all had experienced an ordinary dilemma with feelings of guilt. As they further questioned their assumptions, such as towards the use of water, in their everyday life, they showed a bit more critical self-assessment, although whether transformation took place or was taking place still requires further exploration. This feeling of guilt is similar to previous studies that they have imaginaries of the Third World countries or the Global South in scarcity of resources, except that these imaginaries or narratives came from previous volunteers. Fairly speaking, the conditions in the village the previous project team experienced were less pleasant, which explains the pre-trip perspectives constructed for the new project team.

Many of the volunteers said that they wanted to get immersed into the local lives and have an experience of doing what the locals usually do, through which they could learn and understand some local cultures and practices. Particularly in the village, they sat on the floor to eat even though ants were crawling around, or as what Gwendolyn said, ‘if we are in India now, when they eat with hands, I definitely won't ask for chopsticks… what they do, how they do it, I will do the same, I won't do it in my own way’. However, their practices just could not be changed in a very short period of time because they had a certain way for doing things from their habitus. Ona recalled the guilt she felt once while they were cooking. She said,

> they [the villagers] still ate the rotten potatoes but we didn't. We threw the vegetables away if the leaves got slightly yellowish or what, sometimes when they saw us cooking, they felt strange as we threw away things that were edible to them, while Hong Kong people thought they were not edible.

This was the usual way of handling rotten potatoes at home, which did not make them feel guilty about wasting food before they went to Cambodia. However, this sense of guilt emerged when they saw how the locals were doing it, and it got stronger when the locals were around, which also made them worried about how they would appear to the locals. Same as using water, this was so different from how they would do it in Hong Kong that they were put into a situation of re-examining their practices from the convictional and behavioural dimensions. The volunteers were bound by criteria of moral judgement which were already in place from their everyday lives and also within their moral community (discussed in Chapter 3), which led them to evaluate what they were doing. Their ordinary ethics was then challenged by another set of moral codes in this ‘other’ place of different social and cultural contexts. Ona was more concerned about the amount of water they used and the waste they created in relation to the local lifestyle and the actual scarcity of resources, while Selina did not feel the same guilt because she was more focused on the behavioural change according to the context, resulting in a different intensity of guilt. Either way, transformation could be facilitated or effected through dialogue with others or encountering a different culture, where the 'unfamiliar' experience helped to challenge the 'familiar' everyday situations and practices, such as showering and dealing with rotten vegetables in these examples. The emotional responses, such as guilt and shame, lead to a questioning of the taken-for-granted frames of reference.

Another form of experiencing dilemma was observed in demonstrating an ethical narrative. Bungalows elevated above the river and surrounded by mangroves in the Trapeang Sangke Community, which cannot be seen in Hong Kong the concrete jungle. Staying in such an environment on a service trip made them think about the nature of a service trip and the practices and image of volunteers. Because of the nice environment per se, or because of the guilt of staying in such a nice environment as a volunteer, it seems that the volunteers tended to put an 'ethics tag' on their experience and wanted this to be seen explicitly and publicly. One volunteer posted on Instagram a photo of herself lying in a hammock with the surrounding environment in the background and a caption of 'service trip', and another photo showing the darkened soles of her feet again with the sunset, river and bungalows in the backdrop and the caption 'these dirty feet dayzzzzZZZzz'. The second photo's caption encapsulated both her hard work of getting into the mud resulting in the dirty feet and a feeling of relaxation from the scenic backdrop and her twist of 'zzzzZZZzz'. She captioned the photo with 'service trip' partly because she received messages asking if that was really a 'service trip' after her post of sunbathing in a hammock. This could be an internal self-examination with feelings of guilt, as this lack of shock was indeed challenging their assumptions. This dilemma and the sense of guilt came from the different assumptions of a service trip between her friends and herself. It also came from what good qualities of a volunteer tourist demonstrate or accumulate in such context. This then questions what dilemma and whose dilemma is disoriented. This 'ethics tag' was a way to shift the focus to highlight what they were doing was tough work despite the bungalows and river view, hoping to tell the right story or make the narrative appropriate, which is different from telling the story to the right audience, or to the moral community.

The unfamiliar and the extraordinary

> I think, we know ourselves the best if we are happy or not, whether it's the way of living we like, whether we want to make any changes… volunteering allows us to think from different perspectives and re-adjust ourselves, maybe you can find the true self, find the way you feel happy and comfortable.

Indeed, the experience was remarkable not only because they were put in a place where things were insufficient, but also because what was seen extraordinary from their everyday life were ordinary in the village when their everyday routine was disrupted or suspended. It was also because they were being there, seeing and feeling what was written in books or travel blogs in the same or different ways as portrayed. Some volunteers said these gave them new insights and possibly changed perspectives. This change of physical space outside their realm of everyday life could open up modes of being in the world (Mezirow, 1997). Quinn explained,

> what I value …as you won't shower outdoor in Hong Kong, is that it's often that you got the chance to gather to cook together, and wash up together, to shower in groups, living that can only be experienced there.

This was the experience of being-with a group of volunteers free from everyday persona as well as a taste of local ways of living; they have seen some changes negotiated by others in the liminal space of volunteer tourism (discussed in Chapter 3). It was also the exploration of different ways of being by slowing down the pace of life and going back to the basics. These enabled the volunteers to enjoy lives in the simplest form compared to their busy and competitive everyday. In the village, some of the TBW Team-A volunteers went kayaking in their free time between tasks during the day and stargazing at night. They also did things that they could not or would not do back home, such as taking a shower outdoor, cooking for 10 people or just idling by the river in an afternoon. These were the sort of activities that made no contribution to the local community but just for fun. However, the connections among tourism, fun and learning should not have been neglected (Falk, Ballantyne, Packer, & Benckendorff, 2012). Sin, Oakes, and Mostafanezhad (2015) have also reminded us, it is almost impossible to separate or extract pleasure from the analysis of actual volunteer tourism experiences, as it is clearly highly aesthetic and not pleasure-free when viewed as a form of ethical tourism. These terms should not be seen as mutually exclusive, although they have been treated as such, in order to reconcile this responsibility and pleasure, ethics and aesthetics. Further from this, this constitutes a specific geography that is conducive for changes through tourism and learning where we do not see volunteer tourists only as the ethical global citizens who are always responsible, compassionate and committed. In this context, it was the leisure or idle time when the volunteers critically reflected on their lifestyles by going through the process of negotiating their goals, values, feelings and meanings of life. For example, Gloria talked about how she searched for meaning and purpose of life by taking up new challenges to explore possibilities, chatting with others, both project team members and locals, and introspecting during her own time. We had several deep conversations about lifestyles in different countries, our attitudes towards stereotypes, and how to strike a balance between expectations and self-satisfaction. Once on the coach, she shared,

> I had given myself a lot of time to calm down during the trip, I felt like I was living in the moment. Although I haven't figured out what I want to

> do or how I want myself to look like, I've got the time to ponder and gather my thoughts.

She said later after the trip,

> … I felt like their life is very leisurely, it seems they just do what they need to do every day, or they concentrate on only one thing every day, for instance we saw an elderly woman peeling the shells off peanuts for sowing, she was still doing it on our way back, she was focusing on doing that one thing.

Gloria was jealous of the elderly, or the local people in general, that they were leading a simple lifestyle, which was exactly missing in volunteers' daily life so that they quested for, as many volunteers shared similar thoughts. Undeniably, there could be embedded assumptions of the volunteers that had ignored the social complexities of such scene like previous studies critiqued (e.g. Baillie Smith, Laurie, Hopkins, & Olson, 2013; Mostafanezhad, 2014). However, I would like to look at how this particular context led to the volunteers' discontent or led them to seek content in contrast to their everyday lives. The rhythms and temporal suspension of the everyday moral codes helped to put a pause to all sorts of stress, competition, frustrations and other problems in this liminal space. This temporary escape from their fast-paced everyday also allowed them to enjoy the tranquillity of the natural environment and the slow-paced lifestyle of the villagers, which made them think about what they used to do back home, what the purpose of doing those things was, and what they could really feel from their fast-paced lifestyle. Observing how local people led their life or learning how the villagers in the Trapeang Sangkae Community joined together to restore their fishing grounds, volunteers reflected on how much happiness and enjoyment they could derive from this so-called very simple lifestyle, from which they were trying to find a different pace of life, even though sometimes they were unaware of this process of critical reflection. As the difference or contrast becomes greater and more apparent, the conscious negotiation of the different temporalities that constitute their everyday life led them to reflect on and possibly seek alternative practices in their daily lives (Fullagar, Markwell, & Wilson, 2012). Appreciating and embracing a slow life as a lived experience in their otherwise 'fast world' opens up ways of being through re-conceptualisation of time in their everyday life. This interruption of the pace of the 'fast world' allows them to reflect and re-invent themselves during the course of an ethical action (volunteering in this context), possibly leading to plans of action for change if the 'exploration of options for new roles, relationships, and actions' (Mezirow, 2000b) has been positive and productive. This also speaks to the ethics of care for the self and developing the moral self through leading a meaningful life.

The change of environment provided a great intensity of encounters that resulted in disorienting dilemmas and critical self-reflections. This implies a unique and interesting geography of where and how disorientation and critical reflection take place. Crossley (2012) states that "moral redemption is performed by volunteer tourists as a spatial practice, involving travel to spaces charged with enough potency, in terms of extremes of poverty, to bring about the transition to an 'appreciative' state of mind" (p. 244). The feelings of guilt or even shame partly came from the privilege of the volunteers and the presentation of the self in front of the locals, which is similar to what has been largely discussed in the literature (e.g. Crossley, 2016a; Crossley, 2016b; Simpson, 2004). Nevertheless, such encounters could also be more intense in stimulating the changes in the mindsets of the volunteers because these volunteers were not there to see how bad the living conditions in Cambodia were, or how grim and miserable the lives of local people were, as other forms of alternative tourism such as slum tourism usually aim to do. Nor were they there to observe poverty and make them feel 'more fortunate' and more appreciative of what they have at home (Simpson, 2004). The 'appreciation' of the Hong Kong and Taiwanese volunteers shown in this and previous chapters was not the same 'ethical end' which was suggested by Crossley (2012), that allowed the Western volunteer tourists in her study to return home knowing that they had undergone a personal transformation but with the same lifestyle. Although the volunteers' lifestyles in this study were challenged by the local way of dealing with rotten vegetables or slow living, this challenge and thus the 'us and them' binary was not based solely on the social inequality or comparison of material differences. It was also due to the way of doing things and the relative moral codes in this liminal space, which created a sense of guilt and shame that was different from the kind of 'appreciation' or even guilt-free experience of previous studies (e.g. Crossley, 2012; Simpson, 2004). Thus, it was the travel from the familiar to the unfamiliar, from the hustle and bustle to the calm and simple, or from the highly socially alienated to the deeply engaged and connected that shocked or awakened the volunteers, making them more conscious of different dimensions of their lives that had long been ignored or repressed in their daily lives. These encounters led them to have a deeper understanding of the self, to question their beliefs and ideologies, and to reflect on their everyday behaviours and lifestyles.

Transformation on the move

> From my experiences, the change happens during the trip, but it extends beyond the trip. I think the change is not like a straight line, it is rather fluctuating, it's a shock from time to time. (Dawn, Taiwanese volunteer)

According to Mezirow's 10-phase of transformative learning, it comes to devising plans of action and trying out new roles in order to reintegrate into "one's life on the basis of conditions dictated by one's perspective". Coghlan and Gooch (2011) found that existing volunteer tourism programmes lack the two steps of developing a plan of action and trying out new roles. Given the fact that many of these service trips are very short, I suggest that the steps in the transformative learning theory could be understood in a non-linear sense that the moral fulfilment gained from the belonging to moral communities may help the volunteers to reintegrate into their ordinary lives with new perspectives or disrupted frames of reference, and then to further explore and plan for actions. The volunteering space serves as a space for critical self-reflection, either because the volunteers are distanced from their everyday life or they are exposed to shock, or both. During the trip, different activities acted as platforms for searching the possibilities and the experience from different encounters was intense. The moral communities and communitas developed facilitated and encouraged the opening up of the self and thus of possibilities or perspectives, making the feelings more intense and heightened. These feelings may come at different times or with different intensities; the volunteers believed that such feelings or emotions from the experience would not fade away when the programme ended. Here, volunteer tourism is not just about physically being there with the intention of acting morally, as discussed in previous studies. It is also about the intensive immersion into a new frame of references in the liminal space, and the experience is a catalyst for transformation. The process may involve a prolonged period of confusion and muddling, but the activities during the trip may trigger a breakthrough or a sense of enlightenment (Cranton, 1994), during and/or after the trip. For example, Dawn explained,

> the shock at the moment of happening should be the greatest, and its intensity would be lower after returning home… but it won't fade away… I can still remember those feelings, they won't make me feel that I need to do something immediately. I am still thinking what I can do, in what ways, but I won't forget as I have written it down, also the feelings are so profound that they can hardly be forgotten. I have got many inspirations for my life direction and goals, but it's not immediate, it may be immediate for some people like [those who founded] Light Love… It is a catalyst in our daily life, you can have different perspectives when seeing the same thing.

These trips that last for less than four weeks would be categorised as 'shallow' in a conceptual framework proposed by Callanan and Thomas (2005). However, for the transformational potentiality, I suggest that we should look at the temporality of the 'shock' from the encounters. If the intensity of the encounters was high enough, the experience could last as a memory or a reminder, as a sense of being in their daily lives. This reminder was somehow sustained by the communitas which did not fade away after the trip but was lingering in its own way over time. This communitas also involves *qing*, the relations or emotional connections with people they know, with whom they did something together.

Travelling from the familiar everyday to a new environment, the observations and contact with local people could lead volunteers to introspection. As experienced or heard from the volunteers, most of the changes took place after the trip, particularly in the behavioural dimension. Their actions and new roles could take the forms of continuing the act of doing good for the locals, bridging some ethical practices back home, and changing lifestyles and views through reflecting on global issues. The founding of Light Love mentioned by Dawn was one of the examples. Three teenage founders of Light Love were in Siem Reap for their service project at the same time as our ELIV group, so they gave a talk on how they started the project. During their service trip to Cambodia in 2012 also organised by ELIV, they noticed that in the village they visited there was a need for lighting at night especially for women going outdoor to use the 'natural toilet'. They later gathered to discuss what and how they could do something for the villagers, so they came up with the idea of sending solar-powered lanterns to rural families in Cambodia. They raised funds through their schools and parents' networks as well as applying for project funding in Taiwan, and have made various trips during their school breaks to send lanterns to thousands of families since 2015.

Similarly, Camellia in the Taiwanese group was mulling over the idea of setting up a sale platform for the palm sugar and other products produced by the villagers. She shared,

> … I have been looking for what I can do, what I am able to do…. In the past, I would have just given up as it involved laws and regulations, but now I checked the relevant law and regulations. That's the change I didn't expect on myself. I want to try, to see how far my ability can help. I am still at the stage of gathering information as I work full-time… see how much I can do…Even my boyfriend said I have become another person after coming back…My enthusiasm usually did not last long, but this time I have a connection with Cambodia, so I want to give back as they have given me a lot.

Actualising this plan takes time in understanding the regulations, devising sales and marketing models, and reaching a consensus with the villagers. This was not implemented in the way she wanted, also due to her other personal reasons. This did not result in a positive impact on the villagers who she has never met. But the process uncovered a new dimension or quality of the self, led her to a different attitude towards accomplishing a task, and exemplified that she took actions in achieving the plan. These actions included proposing and discussing the idea with Jenny (the local Project Manager) and checking the regulations. These echo the three dimensions of changes as well as steps five to seven in Mezirow's transformative learning theory. Camellia's changes or plans of doing something further shows that the transformation was being carried on after the trip, although the plan was not actualised, the psychological and convictional changes here last for longer, or have been used in other aspects.

There were also times when the volunteers picked up issues from discussions and observations, they have showed attempts to bridge them back to their everyday lives. This may happen at the time of returning home, and it may be either the same practice that is aggregated back from the liminal space, or behavioural changes at the stage of aggregation that have informed different ways of living. For example, after discussing the issues facing the fishing community, such as the loss of fisheries due to the deforestation of mangrove forests for development, the global cycle of waste, and general environmental degradation in a vicious circle, Kathleen had a heightened awareness and began to relate issues from her home country to the global, and from herself to people living in another place. Then she started making some changes in her daily life in the hope of influencing those around her as well. She shared,

> those issues are actually related to us. The first thing I can think of is that many big corporations always set up factories in those countries like Cambodia and Vietnam because there are fewer regulations. Why don't they set up in their home country? Because of NIMBY .... Then why you can put something you don't want in other's home? ... Why you have sufficient supply of water, because you use their water. Why I feel happy or things are normal because my waste is disposed to others' place... I think, that's not okay. Sadly, many people in Hong Kong would ask 'what bothers you?'...

She continued,

> I turn off all the lights at home, whereas my mum usually leaves one before going out, but I will turn it off... she understands that I try to save resources, not just to save money, but it's not okay if things get worse. I am more aware of environmental issues.

Again, whether she would keep doing these for one month or one year and more is still unanswered here. But what is worth attention to here is the prolonged process of transformation beyond the service trip rather than transformation at the end of the trip as an outcome of the volunteer tourism experience. It is also because the intensity of those experiences was somehow sustained by the relational ethics which connects her and Cambodia, and the inspirations she got to reflect more critically and seriously on some global issues that have been discussed regularly.

In contrast, Amelie felt hat she was still in the process of exploring, an extended process from her first volunteering experience in Taiwan with ELIV two years earlier. She said,

> at the end of that trip, our team leader gave each of us a letter, in which he asked me to join Project Cambodia ... he said when he saw me, I looked like I was in search for something, so he recommended me to join the Project. He found himself in Cambodia, so he hoped that I could find what I was looking for after being there.

Amelie and her mother had opposing views on how her life should be run, showing a clash between old and new ethics (see Chapter 2). Having followed her mother's wish to complete a Master's degree, she wanted to have some kind of freedom to lead her own life and make individual decisions. The letter thus gave her a push to volunteer in Cambodia. Through interactions with other volunteers and local people, observing and engaging in different lifestyles, and learning about others' difficulties and challenges in Cambodia, she said she 'got inspirations from this land'. Besides signing up as an ELIV team leader after our trip, she returned to volunteer in Qinghai where she had first volunteered overseas, hoping to reconnect the self with her original passion. While she was trying new roles or actions, at the same time she was going through again initial steps of transformative learning to search the desired self through ongoing, deeper understanding of the self.

These occurred at the aggregation stage from the liminal space when volunteers returned with a new status or state of mind undergoing transformation or preparing for transformation. The process of critical self-reflection may even persist longer than the trip itself, postponing the phases of planning of actions and trying new roles to the aggregation stage, or even re-starting the earlier phases of transformative learning before they could know what actions they need to take. This is dependent on the multiple subjectivities of the volunteers in the particular context of social group experience, relationship and life course, and thus how the transition from volunteering back to everyday life takes place (Baillie Smith et al., 2013; Hopkins, Olson, Baillie Smith, & Laurie, 2015). This is similar to the transition from volunteer to activist found by McGehee and Santos (2005), as well as the challenge of renegotiating the self on return faced by the young people in the study by Hopkins et al. (2015). Although what the volunteers would do after returning from the trip may not be directly related to the good cause of the volunteering projects such as environmental issues, education, domestic violence and street kids, their enhanced awareness was embedded in their plans of taking actions or new roles, or changes in practices. These are processes of shaping and re-shaping the self while searching for different dimensions of the self, through which it is adding values to the individual from the learning. However, this spatio-temporal dimension of transition, while significant, still requires more attention in the literature.

Undeniably, these service trips are short-term by nature; but here we see the potential of volunteer tourism to develop the moral self elaborated from the practices of the trip into a more sustainable one, or into a wider scope in post-trip everyday life. Many volunteers mentioned that 'you cannot pretend not to know after you have known about it'. Various activities and discussions helped them to learn different issues and cultures, and then to reflect and rethink their roles and responsibilities, their relations with others, and the nature and causes of problems as part of a community and the world, when people and things are getting more globally interconnected but at the same time socially alienated. The experience bears the potential to make them become more conscious and to drive them to take action. Having been the moral subjects created, they have their assumed responsibility, both for themselves and for others. If the consciousness and catalysts to change last, these identities can be translated into moral capital and authentic, ordinary qualities of the moral subjects over time, rather than only performing and acting to reinforce the moral identities. Once internalised through the experience of being-with, being-alongside and being-in, these moral subjects could become the individuals who are more conscious of what they do and what they can do in their everyday lives, further developing the self. Admittedly, how much quality is added to the body is questionable, or unable to measure objectively, as the inculcation or accumulation is sometimes unconscious or implicit and it can take various forms. This would require efforts from sending organisations or the sector of volunteer tourism to facilitate the process of transition as well as that of developing the self.

References

Baillie Smith, M., Laurie, N., Hopkins, P., & Olson, E. (2013). International volunteering, faith and subjectivity: Negotiating cosmopolitanism, citizenship and development. *Geoforum*, *45*, 126–135.

Bamber, P., & Hankin, L. (2011). Transformative learning through service-learning: No passport required. *Education & Training*, *53*(2/3), 190–206. doi:10.1108/00400911111115726

Callanan, M., & Thomas, S. (2005). Volunteer tourism: Deconstructing volunteer activities within a dynamic environment. In M. Novelli (Ed.), *Niche tourism* (pp. 183–200). Oxford: Butterworth-Heinemann.

Coghlan, A., & Gooch, M. (2011). Applying a transformative learning framework to volunteer tourism. *Journal of Sustainable Tourism*, *19*(6), 713–728. doi:10.1080/09669582.2010.542246

Cranton, P. (1994). *Understanding and promoting transformative learning: A guide for educators of adults*. San Francisco: Jossey-Bass.

Crossley, É. (2012). Poor but happy: Volunteer tourists' encounters with poverty. *Tourism Geographies*, *14*(2), 235–253. doi:10.1080/14616688.2011.611165

Crossley, É. (2016a). Affect and moral transformations in young volunteer tourists. In *Emotion in motion* (pp. 85–97). Oxon: Routledge.

Crossley, É. (2016b). Poor but happy: Volunteer tourists' encounters with poverty. In *Tourism and geographies of inequality* (pp. 41–59). Oxon: Routledge.

Falk, J. H., Ballantyne, R., Packer, J., & Benckendorff, P. (2012). Travel and learning: A neglected tourism research area. *Annals of Tourism Research*, *39*(2), 908–927.

Fullagar, S., Markwell, K., & Wilson, E. (2012). *Slow tourism: Experiences and mobilities*. Bristol: Channel View Publications.

Hammersley, L. A. (2014). Volunteer tourism: Building effective relationships of understanding. *Journal of Sustainable Tourism*, *22*(6), 855–873.

Hopkins, P., Olson, E., Baillie Smith, M., & Laurie, N. (2015). Transitions to religious adulthood: Relational geographies of youth, religion and international volunteering. *Transactions of the Institute of British Geographers*, *40*(3), 387–398.

Knollenberg, W., McGehee, N. G., Boley, B. B., & Clemmons, D. (2014). Motivation-based transformative learning and potential volunteer tourists: Facilitating more sustainable outcomes. *Journal of Sustainable Tourism*, *22*(6), 922–941. doi:10.1080/09669582.2014.902065

McGehee, N. G., & Santos, C. A. (2005). Social change, discourse and volunteer tourism. *Annals of Tourism Research*, *32*(3), 760–779. doi:10.1016/j.annals.2004.12.002

Mezirow, J. (1991). *Transformative dimensions of adult learning*. San Francisco, CA: Jossey-Bass.

Mezirow, J. (1996). Contemporary paradigms of learning. *Adult Education Quarterly, 46*(3), 158–172.

Mezirow, J. (1997). Transformative learning: Theory to practice. *New Directions for Adult and Continuing Education, 1997* (74), 5–12. doi:10.1002/ace.7401

Mezirow, J. (2000a). *Learning as transformation: Critical perspectives on a theory in progress. The Jossey-Bass higher and adult education series*. San Francisco, CA: Jossey-Bass.

Mezirow, J. (2000b). Learning to think like an adult. In Jack Meziro and Associates (Ed.), *Learning as transformation: Critical perspectives on a theory in progress* (pp. 3–33). San Francisco: Jossey-Bass.

Mezirow, J. (2003). Transformative learning as discourse. *Journal of Transformative Education, 1*(1), 58–63. doi:10.1177/1541344603252172

Mostafanezhad, M. (2014). Volunteer tourism and the popular humanitarian gaze. *Geoforum, 54*, 111–118. doi:10.1016/j.geoforum.2014.04.004

O'Sullivan, E. (2002). The project and vision of transformative education. In E. O'Sullivan, A. Morrell, & M. A. O'Connor (Eds.), *Expanding the boundaries of transformative learning: Essays on theory and praxis* (pp. 1–12). New York: Palgrave Macmillan US.

Simpson, K. (2004). 'Doing development': The gap year, volunteer-tourists and a popular practice of development. *Journal of International Development, 16*(5), 681–692. doi:10.1002/jid.1120

Sin, H. L., Oakes, T., & Mostafanezhad, M. (2015). Traveling for a cause: Critical examinations of volunteer tourism and social justice. *Tourist Studies, 15*(2), 119–131. doi:10.1177/1468797614563380

Taylor, E. W. (2008). Transformative learning theory. *New Directions for Adult and Continuing Education, 2008*(119), 5–15. doi:10.1002/ace.301

Zavitz, K. J., & Butz, D. (2011). Not that alternative: Short-term volunteer tourism at an organic farming project in Costa Rica. *ACME: An International E-Journal for Critical Geographies, 10*(3), 412–441. doi: 10.14288/acme.v10i3.905

5 Disorientation and re-orientation

Reshaping the moral self by sending organisations

As discussed in previous chapters, individuals, particularly in Asian cultures, often refrain from explicitly expressing their feelings or engaging in deep interactions, meaning that their assumptions are less openly challenged. Volunteer tourism provides a unique context, where space, time, and audience enable participants to face disorienting dilemmas and engage in introspection. These opportunities for self-reflection, which arise either naturally through social interactions or are crafted by sending organisations as part of the programme, allow for questioning deeply held values, attitudes, and beliefs. Travelling to unfamiliar environments and interacting with others disrupts the volunteers' meaning perspectives, prompting reflection – whether conscious or unconscious – on their ways of thinking. The role of sending organisations in this process is critical, as they strive to ensure that service trips remain 'ethical' or 'doing good'. These organisations also instrumentalise activities and contexts as tools to foster self-reflection, disorienting dilemmas, and re-orientation. This chapter will explore what roles these organisations play to facilitate critical self-reflection and other transformation through the creation and utilisation of such contexts, and what this implies for the development of the moral self.

Framing of volunteer tourism

The moral communities discussed in Chapter 3 embodied their joint effort and trust. This joint effort was also the language used by the sending organisations to frame volunteer tourism as a form of sharing labour, time and skills. ELIV's Project Cambodia and the mangrove restoration project that TBW Team-A participated in both are long-term projects. At the time I volunteered, Project Cambodia was running for the fourth year in the same village building houses for families with needs and toilets shared by four or five families to improve the sanitation. The mangrove restoration project was initiated by the local fishing community in 2012 to restore their lost mangrove forests. They collaborated with a local NGO to receive international volunteers to support the work. The efforts from each project team were like a relay, a small contribution to the change that the organisations wanted to achieve, at the same time the work was

DOI: 10.4324/9781003413929-5

not only done by the volunteers. For example, the mangrove seeds collected and potted for nursery by one group of volunteers will be monitored by the locals and replanted either by the locals or the next group of volunteers. For the ELVI projects, before building the house, the local project manager asked the family that would receive the house to invite their relatives and friends to take part in the process so that they would not receive the house without any contribution. Same for the toilet, those families that would share it were involved in the construction work. The house and the toilet then were not only give-and-take objects; they symbolised the joint effort, transmitting the message that it is an act of sharing.

Speaking with those images used on the organisation's website, at the information session in the university and pre-trip briefings, staff from both TBW and ELIV highlighted the ability of the volunteers to make a change, or it is 'time for change'. As discussed in Chapter 3, the sending organisations consistently framed the service trip as an act of sharing rather than a form of sending aid. The act of sharing was instrumentalised in different forms, one of which was sharing resources. One of the pre-trip tasks for the ELIV service trip involved collecting items in new or good condition, such as tote bags, clothes, water bottles, and stationery to bring to Cambodia. The idea was to share what we had in excess at home with people (the local school and families involved in building the house and toilet) in Cambodia who would need those items, to reduce waste and give a new use to the items, and to suggest sustainable practices such as substituting plastic bags with reusable bags. To avoid exporting trash to Cambodia, we received a guideline on what were suitable and useful. The first task upon arrival was sorting the materials led by Jenny, the local Project Manager, to make sure the items in each bag were appropriate and similar in quantity and to avoid creating the feeling that the volunteers were just giving them a collection of 'stuff' when the local people received them. When handing the bags to the local people, we were reminded to hold the bag with both hands to show respect, and to squat to greet them at eye level especially handing to the children in the school, a gesture that underscored the interaction as one of equality.

Many volunteers agreed with this framing of volunteer tourism. For instance, Xena said,

> it's not that I am superior and others [locals] inferior, it's an equal relationship… I like the idea of 'sharing'. When I decided to join, I thought I had this ability so I wanted to help, I did have this thought of providing help, but I did not think I am from a higher ground or more superior. It was until he [the program director] mentioned 'sharing' that I thought it's a more appropriate way to frame voluntary service. I think 'help' is not inappropriate, but it may lead us to reflect on the social status of the volunteer and the volunteered. 'Sharing' is more accurate.

While the marketing materials of the sending organisations speak to the ethical dispositions that frame the motivations of these volunteer tourists (see Chapter 2), the discourse of the sending organisations also matters in disorientation and reorientation of dilemmas towards the reshaping of the moral self. Many reported cases in the literature show how the sending organisations may reinforce the North-South divide (Polus, Carr, & Walters, 2023) and the consumption of the local communities for humanitarian fulfilment by projecting an image of villagers being extremely miserable or the volunteers being the saviour (Baillie Smith, Laurie, Hopkins, & Olson, 2013; Bandyopadhyay, 2019; Laurie & Baillie Smith, 2018). Here shows a different case that highlights the relational ethics and the collectivity and mutuality of a community throughout and beyond the experience, which is somehow instrumentalised by the sending organisations.

This framing of volunteering somehow helps to set a lower tone for the service trips and the work done by the volunteers. In traditional values, it is a virtue to be humble rather than making everything public, and it is believed that volunteering should be done in a modest manner. Developing the moral self is a good thing to do, but virtue signalling is not considered good or appropriate in traditional Chinese culture. People prefer not to talk about personal or family matters or are otherwise reticent. Wendy's father did not think volunteering is a bad thing, but he thought volunteering should be done or seen as an ordinary practice. When Wendy was in charge of marketing of volunteering programmes, her father thought her marketing strategies were too explicit and the marketing materials were too high-flown, like how Fair Trade tourism is used by tourists to demonstrate the ethical self (Boluk, 2011). According to her father, doing voluntary work should be low-profile, instead of like making an announcement that 'I go volunteering, it is amazing, it is great'. He saw volunteering as an opportunity given by others to the volunteers to help. While it is the responsibility of sending organisations to prevent creating rhetoric of volunteering as a change-the-world act, or over-mobilising people, it is an ethics of the individuals not to talk 'too loud' about themselves volunteering to avoid virtue signalling. Rather than a sense of entitlement, humbleness in Confucian values comes in to negotiate with the ethics of globalising responsibility and the practice of virtue signalling. As Wendy said,

> a bit earlier, he reminded me that, when others allow me to help them, I should thank them, because they give you the opportunity to help them, or they are willing to be helped, we may think this is good for them and do it on them, but it is not always good, so he thinks I should be thankful if they are willing to give me the chance to try, instead of saying like 'see, I am here to help you'.

Such discourse on sharing and exchange was instilled through pre-trip and post-trip workshops as well as discussion and reflection sessions during the trip. Instilling such idea of sharing might help to negotiate the criticism of why going on volunteering overseas or paying to do hard work from the older generation based on the old ethics that the volunteers mentioned. While the sending organisations created contexts for disorienting dilemma, they also helped to translate self-examinations of guilt and assumptions into new perspectives and values, or even actions. For example, Helen said,

> it helps me to develop myself, to develop myself through different challenges or shocks… when your values are challenged, then you will think about what one is right or wrong, good or bad, through the process of thinking, then you may have a new perspective. Let's say volunteering, after that I have started to think bout which is a better way to help others, a balance, gradually, I may have an answer, while values and perspectives are being shaped.

These settings helped to reorient the volunteers to think beyond the issues that they observed or heard. Rather than seeing those issues as problems to be solved, they were guided to think critically about what they are doing from their roles in their everyday life, what they could do with their lives and who they want to become, like various examples given in the previous chapter. While TBW organised four post-trip workshops and field trips to connect the challenges facing the Muslim minority in Cambodia[1] and those of South and Southeast Asian communities in Hong Kong, ELIV did not have any post-trip event but an organised tangible reminder. In the last reflection session, each of us was asked to write a letter with the subject 'I Promise I Will' to ourselves and it would be posted to us three months after the trip. As what the team leaders believed, it did not matter if the change or plan of change was significant or trivial, what mattered was the act of making a resolution and taking action. For example, Amelie shared that her promise was to develop a habit of using reusable cutlery. Sustainable development is one of the causes underlined across ELIV's projects and specific activities. In order to reduce plastic and food waste, we had to bring our own reusable container and cutlery for lunch in the village and to get our portion from the big container. Amelie had been bringing reusable cutlery when eating out already before the service trip, but only occasionally. After the trip, she researched different reusable substitutes and put them into practice to make the change she had promised, regularly sharing with the group what she had found. Together with the intangible reminder or memory from the communitas developed, the letter she received three months after the trip also served as a physical reminder of her planned action.

Travelling not like a tourist

> Before going to the village, we went to S-21 and Killing Field which were the two places I wanted to go to know about their history, plus there was nothing I wanted to buy, so I didn't really plan to be a tourist.
>
> (Selina, Hong Kong volunteer)

Visits to various tourist attractions, including both the must-sees and less touristic ones were charged with a theme. Before going to the villages, TBW organised a tour to S-21 and Killing Field, the sites of the Cambodian genocide during the Khmer Rouge Regime, now converted into museums. The tour was aimed to provide some background knowledge of the country and to stimulate dialogues about the genocide, the political system, social issues and the affects and emotions that remain from the Pol Pot era. While we were visiting the museums like other tourists, what made it different from package tours or independent travel was the debriefing. This also provided the volunteers with more understanding, and topics to discuss with locals such as the hostel staff and tuk-tuk drivers.

After the activities in the school, ELIV organised an outing with the school children to the Angkor, the most well-known UNESCO attraction. Although it is free for local residents to go to the Angkor, not many children in the village have visited there because it is too far from the village or their parents do not have time. Each volunteer was paired up with one or two children, they were given a guided tour in both Khmer and Mandarin, and did some sketching of the temples. They were toured around the temples like other tourists, but what made it different from a typical sightseeing was them accompanying the children to appreciate their heritage, the time being with or only being alongside the children given their language barrier.

Gwendolyn said, her experience with her accompanying kid was more prominent than that of seeing the Angkor. She said,

> I may not specifically remember the Angkor, I would remember the Angkor mainly because of the time I spent with the kid there and how the interaction was like, instead of telling you about specific attractions or photos I took.

Of course they would remember the Angkor from the photos they took. But their visit to the most touristic attraction in Cambodia became more significant and unique because the outing was designed in a way that they saw the temples and things with and through the children, from their sketches and keeping them company. The experience was particularly prominent for her as a teacher of special education needs, the contact with these village children made her think about what she could do more for her students with different issues of accessibility. Education was one of the topics of dialogues among the team; the school activities and outing undeniably stimulated some discussions and reflections. It, however, might rationalise the purpose of the outing, taking photos of the children, and using them to tell the story.

Visiting the Pub Street in Siem Reap city centre, the most crowded and touristic place of the city, was another scheduled activity by ELIV. It was not the usual free-time activity after dinner; rather, it was a task of observing different groups of people in the street, trying to pick up issues for discussion later on. At the briefing, the Project Director told the volunteers that he had seen several times children carrying their younger siblings asking for money from foreign tourists in the street. He explained that those street kids were controlled by an organised gang using drugs, so the kids had to return with money in exchange for injection. This made a few volunteers want to see the same scene, but to their disappointment, they did not see any. By telling them in the briefing, the volunteers were expecting an intense encounter that normal tourists might not know the story behind even if they had seen it.

Although they did not see any street kids, all volunteers managed to see the victims of landmine explosion performing in the Pub Street and had a short conversation with the policemen on duty. The observation did not end in the city centre. After returning to the hotel, the Project Director led the debriefing on the observations. This debriefing helped to further unpack some of the issues with the support of the Project Director's 10 years of experience in Cambodia. Taking place in the middle of the trip, this debriefing also served as a critical reflection on the roles and responsibilities of volunteers on such issues. The focus was not on their roles in resolving those problems, but rather on how they, as a visitor to Cambodia, could try to understand more about the country and how the past has shaped the present Cambodia. This is similar to going to a museum, but the interaction and discussion made the experience more intense, and the sending organisations played a role in instrumentalising the learning component of a service trip.

'Free time' was sometimes ambiguous, especially on ELIV's itinerary. The 'free time' on the itinerary was scheduled for an optional charity circus show by Phare, a local NGO which supports young Cambodian people in developing skills. Before the trip, the team leaders recommended this circus show in the Facebook group. The whole group went. This structured 'free time' by scheduling a charity-related event before the real free time in the city centre had successfully added some values to the experience of the service trip and something to be discussed at the debriefing. In the last evening, the group was given some free time at the night market after dinner. Because of this kind of constructed imagination and anticipation of a service trip from the itinerary and optional charity show for the first 'free time', Camellia was not prepared for such itinerary. She said,

> when we were free to walk around the night market, I was like, oh my god, what am I going to do? Before the trip, I was not expecting free time in night market like being a tourist, that's why I did not bring a lot of cash for things other than the circus.

Without enough cash to spend, she and Hestia decided to go to the 'Made in Cambodia' Market (introduced by the local Project Manager before the observation at the Pub Street) again in the last evening. She did not get prepared to be a tourist there, she went to that charity market because she thought that was a good and 'safe' place to go. Although service trip was regarded as a form of travelling by many informants, it did not occur to her mind that she would have time for shopping or other leisure activities. This was partly orientated by the team leaders before the trip with the ethical framing of a service trip, forging the experiences in the way that volunteers should be seen as, or they believed they were, travellers acting more responsibly than mass tourists.

Eating with a theme

The volunteer tourism space offers contexts and situations for volunteers to encounter and experience dilemma and negotiation of guilt and shame such as the example of dealing with rotten potatoes (see Chapter 4). Sending organisations, at the same time, instrumentalise the contexts for volunteers to experience, and more actively and directly act on their frames of reference. This could also be considered as an attempt to even off a bit of the guilt from using too much water and wasting food by these more ethical choices of eateries. Both organisations assigned a theme to eating out in Cambodia. Where and what the volunteers went for food on a service trip would possibly help them to understand local situations, to bring up issues of what they can or should do, and thus act more responsibly. On arrival in Phnom Penh, the Hong Kong volunteers were divided into groups of five, with each group given US$2.50 for lunch to simulate living on a tight budget, similar to many locals. Taken to the old market, where locals get affordable options, the activity was aimed to give the group an idea of what it would be like to live on the minimum wage of US$153 a month at that time. Being in such context, Adam said,

> the first impression was that the market looked very dirty, flies over the fried fish, the owner did not bother to drive them away, just too many, but I wanted to try, local food, so we sat down and ordered, it was so delicious.

The NGO practitioners were trying to package the itinerary with some authenticity, places the local people would go and food they would eat. The volunteers heard or saw similar depictions on the media or books, as well as discussed about such issues in Cambodia and other developing countries in the pre-trip workshops. The environment was not pleasant and food did not seem very clean with those flies over; but this outing gave them the intense experience of seeing, doing and feeling it, which made it alternative, exclusive and even premium. However, the sending organisation did not intend to tell the volunteers how they could make a change to issues such as poverty through this activity. At the pre-trip workshops, poverty was one of the themes that the group was interested in, and the TBW included a simulation game and reflection. This activity in Phnom Penh connected the experiences in a staged scenario and a real-life setting, and to a certain extent, helped to negotiate some of their assumptions through discussions and reflections at the first debriefing.

With a similar purpose but in a different approach, ELIV did not go entirely Khmer but focused on the good cause of the selected restaurants. Before each project team arrives, the team leader and local project manager search charity or social enterprise restaurants with a vision and philosophy that ELIV supports. For example, we dined at Café Joe to Go, a local NGO that provides art training to former street children. Although we were served with traditional Cambodian dishes, Amok and Lok Lak, the focus was on issues in Cambodia and the work of charity organisations like this one. When we discussed street children problems after the night market observations, this dining experience was brought up to connect local issues and work done by various NGOs. Similarly, we had a dinner in New Leaf Eatery, a social enterprise founded by two former foreign volunteers, which supports education in Siem Reap through 'providing a high quality of dining experience… with a "taste of Cambodia"' (New Leaf Eatery, n.d.). The well-decorated restaurant full of foreigners featured handmade crafts by local artists, which was directed by the team leaders to browse for the cause behind the dining experience. Probably the volunteers felt reassured by the restaurant's charitable mission, though some questioned whether we were 'eating too well', interestingly, no photos from these meals were shared on social media. This possibly reflects an underlying negotiation between the sense of guilt of eating in that pricey restaurant and the ethical justification of eating for a good cause. Dining for a good cause helped to maintain the 'ethics tag' of the service trip.

This negotiation and embodiment of guilt was similar to that of the volunteer who posted photos of sunbathing in the hammock and the dirty feet but with a caption of 'service trip' (see Chapter 4). What made the difference was that the 'ethics tag' of the service trip here was maintained by the sending organisations and these Taiwanese volunteers avoided any possibility of scepticism induced from an explicit 'ethics tag' for this dining experience. Their self-examination at the same time self-justification were implicitly embedded in order not to disrupt the existing ethical narratives. Both negotiation and embodiment of guilt through the 'ethics tag' somehow helped to reinforce their identity of the moral community, and also showed what and how they wanted to be seen, especially because some of them mentioned that they were criticised for paying so much to go to 'dirty places' or to do hard work. Although the themes of these dinners were covered at de-briefing sessions, there might be a risk that these acts of eating ethically tend to be rationalised instead of being encountered naturally.

Reshaping by sending organisations

When the volunteers decided to join the trip, they wanted to take it as a form of travelling on top of volunteering. Therefore, they had expected more contact with local people than other forms of tourism. Meanwhile, their anticipations and actual experiences were disoriented and re-oriented by the sending organisations. The sending organisations tried to distance the volunteers from the touristic spots and veil their experience as a tourist through discussion of issues from their assigned tasks and various intensity of interactions with local people as a volunteer. This, to a certain extent, helped to ease some of the guilt and anxiety of being seen as ordinary tourists, when they were in a touristy site or popular sightseeing destinations.

Nevertheless, when I asked them about the impacts on the local people, most were not very confident about the extent of the positive impacts on community development which was the main focus of both organisations. Most of them thought the impact was trivial, or focused on themselves rather than the locals. This brings us back to the debate about the dichotomy between 'doing something good' and 'having something good done', or the kind of 'dilemma between local and international volunteering, and the need for international volunteers' that Helen and a few others had. As mentioned earlier, the sending organisations emphasised the acting of sharing, how their efforts could contribute to the long-term project in the villages and have an impact on the local community. For instance, on the day when we went to build the toilet, the team leaders showed us other toilets nearby that were built by other teams, how the villagers have maintained them well by decorating the toilets with hanging plants, and the one we could use for the day which also implied that it was still functioning. This helped to relieve the feeling of guilt or showed how the small efforts of each team contributed to the long-term goal of the community development.

Having the feeling of guilt may facilitate the transformation process. The sending organisations' efforts in enforcing the moral identity may disguise the social inequality, which echoes what earlier studies have argued (e.g. Simpson, 2004). However, what was not fully discussed in those earlier studies but raised in more recent ones (e.g. Coghlan & Gooch, 2011; Hammersley, 2014; Henry, 2019; Knollenberg et al., 2014) and explored in this chapter is how the volunteers are engaged with issues of social injustice throughout the volunteering programme especially in pre-trip and post-trip activities. As we can see, the volunteers and the sending organisations were trying to maintain the identity of a volunteer, the anticipated and constructed moral self, as such identity indeed has the potential in facilitating transformation through self-reflection during and after the trip, especially enforced by the sending organisations to be a more conscious tourist. When they found how little they could do to 'give something back', the ideologies of doing good was negotiated and ameliorated by being guided to think from sending help to sharing resources and later to plan a course of actions. Linking this to the promotional materials, the message of 'sharing' should be streamlined. Without the guided reflection, photos used in posters or on websites would reinforce the subjectivities and imaginaries of the so-called Third World, Global South or developing countries. This may run the risk of creating or consolidating the frames of reference that such act of sharing remains an act of sending aid. It is important to see how the representation and reinforcement of the identity of the volunteers embodied from being part of the various moral communities is translated into additional qualities of the bodies, or *suzhi* of volunteers. In other words, it is vital to see how the volunteers are becoming or have become the kind of individuals who are more ethically conscious of their everyday practices beyond the timescales of the service trip.

Note

1 Muslims constitute about 2% of the total population in Cambodia.

References

Baillie Smith, M., Laurie, N., Hopkins, P., & Olson, E. (2013). International volunteering, faith and subjectivity: Negotiating cosmopolitanism, citizenship and development. *Geoforum*, *45*, 126–135.

Bandyopadhyay, R. (2019). Volunteer tourism and "The White Man's Burden": Globalization of suffering, white savior complex, religion and modernity. *Journal of Sustainable Tourism*, *27*(3), 327–343. doi:10.1080/09669582.2019.1578361

Boluk, K. A. (2011). Fair Trade Tourism South Africa: Consumer virtue or moral selving? *Journal of Ecotourism*, *10*(3), 235–249. doi:10.1080/14724049.2011.617451

Coghlan, A., & Gooch, M. (2011). Applying a transformative learning framework to volunteer tourism. *Journal of Sustainable Tourism*, *19*(6), 713–728. doi:10.1080/09669582.2010.542246

Hammersley, L. A. (2014). Volunteer tourism: Building effective relationships of understanding. *Journal of Sustainable Tourism*, *22*(6), 855–873.

Henry, J. (2019). Directions for volunteer tourism and radical pedagogy. *Tourism Geographies*, *21*(4), 561–564. doi:10.1080/14616688.2019.1567577

Knollenberg, W., McGehee, N. G., Boley, B. B., & Clemmons, D. (2014). Motivation-based transformative learning and potential volunteer tourists: Facilitating more sustainable outcomes. *Journal of Sustainable Tourism*, *22*(6), 922–941. doi:10.1080/09669582.2014.902065

Laurie, N., & Baillie Smith, M. (2018). Unsettling geographies of volunteering and development. *Transactions of the Institute of British Geographers*, *43*(1), 95–109. doi:10.1111/tran.12205

New Leaf Eatery. (n.d.). About us. Retrieved from https://newleafeatery.com/about-us/

Polus, R., Carr, N., & Walters, T. (2023). A new framework: Non-Western perspectives on international volunteering. *Annals of Leisure Research*, *26*(5), 701–715. doi:10.1080/11745398.2022.2041447

Simpson, K. (2004). 'Doing development': The gap year, volunteer-tourists and a popular practice of development. *Journal of International Development*, *16*(5), 681–692. doi:10.1002/jid.1120

6 Conclusion

Reconceptualising volunteer tourism with a non-Western perspective

Reconceptualising volunteer tourism in East Asia

In this book, I do not intend to conceive volunteer tourism only as an alternative to mass tourism, or as a tool for development. I have considered it as a social trend, and have situated the conceptualisation of volunteer tourism around the *self*, to understand what it means in self-making. I have also widened the conceptualisation with a non-Western perspective by drawing on the experiences of volunteer tourists travelling from East Asia, specifically Hong Kong and Taiwan. In order to unpack what this social trend tells us, I am reconceptualising volunteer tourism based on the 'what' and 'why' of motivations and outcomes.

Motivations

Taking part in volunteer tourism does not seem to be just a discrete choice of a single volunteer tourist. It is a prevailing choice from the overarching, complex ethical terrain in which an individual chooses to do things based on a particular instilled position. This choice embodies how structures of the social world are incorporated in the body of an individual in certain ways that structure how the individual acts. Neoliberal constructions and responsibilisation through institutionalising service obligation in schools, together with the promotion of the concept of global citizenship, have developed a culture of volunteering. Embedded in the instrumentalised service obligations and felt responsibility to help the distant others is the moral exemplarity. A role model or moral subject is identified more actively and explicitly to represent and display the good practices and to produce and accumulate moral capital on the individuals. This results in both a formalised and felt responsibility and conditioned response to call for ethical actions. Through these strategies, they have been instrumentalised to think about what kind of person they want to become and what they should do to acquire and accumulate the consonant personal qualities. This is somehow informed by the culturally rooted dispositions in developing oneself or cultivating *ren* in order to become a good person. However, there is a generational difference in the way of cultivating *ren* and

DOI: 10.4324/9781003413929-6

defining a good person. This generational difference rests on the clashes between family duty and ethical imperatives to help others, and thus differing perceptions towards helping others locally and globally, based on the traditional particularistic ethics of care. The encounter of dispositions inculcated from school and from family could challenge the existing mode of thought, or negotiate and add another layer of disposition to the habitus to form a new pattern of thought. This leads to the differentiation between old ethics and new ethics. These multiplicities of rationalising travelling as a volunteer have brought the conceptualisation of volunteer tourism beyond an absolute dichotomy between altruism and self-interest which varies from one tourist to another, to the discussion around ways of being and becoming a moral self. Those promotional materials are thus speaking to the need of developing the good self, or *ren*, from these active strategies and dispositions.

By looking at the complex ethical terrain in which individuals from Hong Kong and Taiwan choose to take part in volunteer tourism, it unfolds a different form of responsibilisation that frames the motivation. This study shows that developing a good self is the choice as well as the outcome, which is slightly different from governmental rationalities that responsibilised individuals are given new opportunities to actively participate in actions to 'resolve the kind of issues hitherto held to be the responsibility of authorized governmental agencies' (Barnett, Cloke, Clarke, & Malpass, 2010; Burchell, 1993, p. 276). This leads us to look at how individuals choose to do things from particular dispositions rather than being posited to make discrete choices.

This social trend also manifests a quest for belonging and rebuilding social ties that can be actualised through being part of moral communities emerged from this travelling activity. Community belongingness touches very deep human aspirations to experience a sense of belonging (Mulligan, Steele, Rickards, & Fünfgeld, 2016). Many informants raised dissatisfaction towards the individualistic culture in Hong Kong and Taiwan, resulting in lost social cohesion and social crisis. Their desire for community belongingness implies such loss in their everyday life in the perceived fragmented society. This then leads to a quest for a temporal escape to distance themselves from the realms of everyday life which they thought is full of problems, competitions and conflicts. The volunteer tourism space offers a liminal space in which moral codes are suspended and ordinary rhythms are interrupted, allowing the volunteers to work and learn with others, experience different facets of the self and find a new self through re-evaluating their own values and perspectives. Volunteer tourism as a prevailing choice mirrors the need for a shift in human consciousness, or a shift in progress, towards a proposed transmodern paradigm (Ateljevic, 2009; Ghisi, 2008; Magda, 2008).

Outcomes

There is an array of earlier studies on the impacts and outcomes of volunteer tourism. This book is not to celebrate the positive impacts of accumulating personal qualities or to criticise the consumption of inequalities for self-recognition through taking part in volunteer tourism programmes. It is rather to highlight that developing a good self is the choice; it is also an outcome. This form of travelling is not just about doing something good; it is also concerned with micro-processes of self-fulfilment, self-seeking and self-questioning during which self-forming is facilitated. Here, I approach outcomes from the perspectives of forming of the moral self, social relationships, and shift in consciousness.

During these micro-processes, an individual is crafted into a global citizen, a class of moral agents. These moral subjects are not simply responsible, rational, autonomous neoliberal subjects with choices; they aim to become a good person, a better self, thus taking part in volunteer tourism becomes an appropriate choice when the rhetoric of marketing materials appeals to this quality. As a volunteer tourist, the active doing of belonging engages the moral subject in collectivity and social interactions, from which the volunteers are able to seek self-identity, to care and to be cared. I-Thou dialogical relation is the outcome of the search for community belongingness, over the processes of growing mangrove seedlings, building the house, and revealing the inner self in a collective manner. The material outcomes in the place embody volunteers' intangible emotional attachment to the *land*. A shift in consciousness may be facilitated when the volunteers are brought into contexts that confront with their meaning perspectives, in which they had deeper understanding of the self, questioning their beliefs and ideologies as well as becoming more conscious of dimensions of their life which had long been ignored or suppressed in their daily life. However, I want to emphasise that transformation is not an absolute outcome; the transformative learning or the transformational process is possible or somehow facilitated by the sending organisations.

This further brings out the spatial and temporal dimensions of the outcomes that contribute to the conceptualisation of volunteer tourism. Volunteer tourists have been criticised for not bringing positive or sustainable impacts to the local communities due to their short stay and lack of skills. Looking at a wider spatial-temporal scale allows us to move our situated understanding of volunteer tourism to a broader framework. As commented by the informants, it is different from making a donation because volunteer tourism is not just about physically being there with the intention of acting morally but also the intensive immersion into a new frame of reference in the liminal space, and the experience is a catalyst for transformation.

The travelling experience provides the right rhythms that interrupt the ordinary ones so that critical self-reflection and searching for meanings are enabled in contexts of shock or the moral communities. Activities or contexts are instrumentalised by sending organisations as tools for disorienting dilemma, introspecting, and re-orienting for changes. The intensity of the experience could be so great that the volunteer tourists would take actions or take up new roles in making contribution to the causes in a different manner, as the transformative learning theory suggests (Mezirow, 2000), or evident in previous studies (Tomazos & Murdy, 2023). However, the muddling period could be prolonged at the aggregation stage so that changes may take place in different localities and forms in a different segment of time. The experience should not be perceived only as a tool to create impacts in-situ; it should be conceived as a trigger. The experience does not anchor at a fixed point, place or time. It should encompass the quality and potential to be nurtured, further muddled, take shape and take place. The temporal dimension is particularly significant as the experience of communitas still lingers in its own way as time passes, which helps to sustain the moral communities and thus moral identity and the inculcated qualities which carries the potential in facilitating transformation to take place, while the impacts caused by the transformation could take place in different forms and localities.

From the experiences of the volunteers of this research, it is a process of becoming. This self-formation is not just about acquiring certain sets of skills but also the dispositions and capacities to act morally (Burkitt, 2002). This becoming is a transition from the instrumentalised moral subject to one who is, more conscious and reflective of their ways of living and thus the meaning of life. It moves from a correlational impact – they are a moral subject so they go to ethically charged places – to a causational impact – after they go on the service trip they become the kind of person who are conscious of their everyday practices back home. Therefore, it would be sensible to look at the possible changes beyond the timeframe and context of the volunteering programme. This geography of change should be acknowledged as a shift from learning about global citizenship to learning for global citizenship. This also suggested a reframing of responsibility and ethics of care that changes on the volunteers' everyday practices could potentially contribute to a different cause in the wider global system rather than narrowly to the single development issue of the distant others. This also aligns with Guttentag (2009) that by orientating the purpose of the volunteer tourism projects to educating and enlightening the volunteers, the negative impacts on rationalising poverty and constructing images of the 'others' can be reduced.

One may raise the question of morality of using the less privileged places as a trigger for change or to disorientate dilemma. However, we may want to acknowledge that this form of travel provides a context for articulating the messy ways of defining, being, expressing and transforming the moral self. My argument for this attempt of re-conceptualisation is that the enactment of ethics of care, responsibility and obligation should not be premised on the narrative that the volunteered are those inferior, without agency while the volunteers are travelling to send aid to change those people's lives. Rather, we should focus on developing a moral self through interaction with the place and people encountered with reference to broader global issues instead of being held solely responsible for issues in another locality. We may even go further to conceptualise the relationship as partnership, as a space for personal learning emerged in which new kinds of alliance and relationships can be forged (Schech, Mundkur, Skelton, & Kothari, 2015). Meanwhile, we should note that this requires negotiating effectively through the programme design of sending organisations and appropriate language so as to avoid reifying or reinforcing the geographical imaginaries. The promises they make in their promotional materials should be streamlined with the programme design with meaningful integration instead of purely a marketing strategy. The muddling and reorientation should also be supported back in classrooms when service obligation is institutionalised and global citizenship is promoted in formal education. These works are necessary to reduce the possibility of perpetual liminality of the volunteers although the transformational process is individualised (Tomazos & Murdy, 2023).

Navigating further

By conceiving volunteer tourism as a social trend, this book has offered an analytical framework to understand how a good self is cultivated, experienced and re-invented through volunteer tourism. It draws together the concepts to examine the active and conscious technologies of the self, unconscious dispositions and the notion of quality which play out for the development of a moral subject. It also links up the concepts of community, liminality and geographies of care to investigate the meaning and formation of the moral self in a collective setting. It then employs the transformative learning theory to discuss the transformation potential of volunteer tourism. This responds to the call for additional theoretical exploration by posing broader theoretical questions and drawing from different disciplines (Wearing & McGehee, 2013). This study has opened up the discussion with the aim of offering some insights for reconceptualising volunteer tourism.

Following these lines, other interdisciplinary interventions would largely contribute to the literature to approach and address growing complexities of global challenges and trends and see where volunteer tourism stands within the global development landscape. One possibility lies in the connections between responsibility, transformative learning and regenerative tourism. Ateljevic (2020) has observed that conscious citizens are acting on transformation towards the regenerative economic systems following a positive transmodern paradigm shift. Röntynen and Tunkkari-Eskelinen (2024) have also proposed a theoretical exploration of regenerative tourism and community-based tourism as guiding principles of volunteer tourism to direct the impacts of volunteers' inputs to the sustainable future of the community. This requires empirical study to explore the role of volunteers play in addition to other stakeholders in such forms of tourism and how this moves the conceptualisation of volunteer tourism further. It also leads to another line of inquiry that extends the understanding the relations of the self with the non-human others in the responsibility dimension of regenerative sustainability.

This book further contributes to the volunteer tourism literature by widening the discussion to a non-Western context with an Asian case study. It is the first volume of ethnographic research of Asian perspectives and experiences of volunteer tourism which has been a field of study premised on Western participants and weighted with Western assumptions and ethical models. While there are ongoing acknowledgement of the Western-dominant conceptualisation of volunteer tourism (e.g. Polus, Carr, & Walters, 2023a, 2023b) and advocates for more inclusive and sensitive scholarship in tourism (e.g. Chang, 2015, 2019; Porananond & King, 2014; Sin, Mostafanezhad, & Cheer, 2021), empirical entries from a non-Western context to the knowledge base is still limited over the past three decades. The narratives and understanding presented in this book are not completely different from that viewing the Western volunteer tourists, but it indicates the importance of recognising the nuanced differences and similarities and adding a multi-cultural perspective to the understanding and conceptualisation. However, this book only presents an ethnographic study of two similar societies in the culturally diverse Asia, not to mention the many other value systems and social complexities that frame volunteer tourism in multiple ways. Take my recent long ethnographic fieldwork (not on the topic of volunteer tourism) in Indonesia as an example. Volunteering is very active across the country, both international and domestic volunteering. Especially during the pandemic, many local groups initiated community development and conservation projects. As told by my field assistant who is an active volunteer on the island, one of the reasons was that they were free. But this probes me to wonder how their initiatives were informed by their traditional values such as *Gotong Royong* (collective action), or how different values interacted when domestic volunteers travelled to another island of different ethnic and religious backgrounds to volunteer in such initiatives. More contributions are needed to pull out the differences and similarities in the hope of stimulating a more just and productive discussion and knowledge production process.

References

Ateljevic, I. (2009). Transmodernity: Remaking our (tourism) world. In J. Tribe (Ed.), *Philosophical issues in tourism* (pp. 278–300). Bristol: Channel View Publications.

Ateljevic, I. (2020). Transforming the (tourism) world for good and (re)generating the potential 'new normal'. *Tourism Geographies*, *22*(3), 467–475. doi:10.1080/14616688.2020.1759134

Barnett, C., Cloke, P., Clarke, N., & Malpass, A. (Eds.). (2010). *Globalizing responsibility: The political rationalities of ethical consumption*. Chichester, West Sussex: Wiley-Blackwell.

Burchell, G. (1993). Liberal government and techniques of the self. *Economy and Society*, *22*(3), 267–282. doi:10.1080/03085149300000018

Burkitt, I. (2002). Technologies of the self: Habitus and capacities. *Journal for the Theory of Social Behaviour*, *32*(2), 219–237.

Chang, T. C. (2015). The Asian wave and critical tourism scholarship. *International Journal of Asia-Pacific Studies*, *11*.

Chang, T. C. (2019). 'Asianizing the field': Questioning critical tourism studies in Asia. *Tourism Geographies*, *23*(4), 725–742. doi:10.1080/14616688.2019.1674370

Ghisi, M. L. (2008). *The knowledge society: A breackthrough toward genuine sustainability*. Cochin, India: Arunachala Press.

Guttentag, D. A. (2009). The possible negative impacts of volunteer tourism. *International Journal of Tourism Research, 11*(6), 537–551. doi:10.1002/jtr.727

Magda, R. M. R. (2008). Globalization as transmodern totality. In R. M. R. Magda (Ed.), *Transmodernidad.* Barcelona: Anthropos.

Mezirow, J. (2000). Learning to think like an adult. In Jack Meziro and Associates (Ed.), *Learning as transformation: Critical perspectives on a theory in progress* (pp. 3–33). San Francisco: Jossey-Bass.

Mulligan, M., Steele, W., Rickards, L., & Fünfgeld, H. (2016). Keywords in planning: What do we mean by 'community resilience'? *International Planning Studies, 21*(4), 348–361. doi:10.1080/13563475.2016.1155974

Polus, R., Carr, N., & Walters, T. (2023a). Challenging the Eurocentrism in volunteering. *World Leisure Journal, 65*(1), 101–118. doi:10.1080/16078055.2022.2146741

Polus, R., Carr, N., & Walters, T. (2023b). A new framework: Non-Western perspectives on international volunteering. *Annals of Leisure Research, 26*(5), 701–715. doi:10.1080/11745398.2022.2041447

Porananond, P., & King, V. T. (Eds.). (2014). *Rethinking asian tourism: Culture, encounters and local response.* Newcastle upon Tyne: Cambridge Scholars Pub.

Röntynen, R., & Tunkkari-Eskelinen, M. (2024). *Voluntourism in the context of community-based tourism, and regenerative tourism: A theoretical exploration focusing on responsibility.* Paper presented at the *Proceedings of the 7th international conference on tourism research.*

Schech, S., Mundkur, A., Skelton, T., & Kothari, U. (2015). New spaces of development partnership: Rethinking international volunteering. *Progress in Development Studies, 15*(4), 358–370. doi:10.1177/1464993415592750

Sin, H. L., Mostafanezhad, M., & Cheer, J. M. (2021). Tourism geographies in the 'Asian Century'. *Tourism Geographies, 23*(4), 649–658.

Tomazos, K., & Murdy, S. (2023). The limits of transformation: Exploring liminality and the volunteer tourists' limbo. *Tourism Recreation Research, 49*(6), 1209–1221. doi:10.1080/02508281.2022.2162695

Wearing, S., & McGehee, N. G. (2013). Volunteer tourism: A review. *Tourism Management, 38*, 120–130. doi:10.1016/j.tourman.2013.03.002

Index

Pages followed by "n" refer to notes.

For Product Safety Concerns and Information please contact our EU representative GPSR@taylorandfrancis.com Taylor & Francis Verlag GmbH, Kaufingerstraße 24, 80331 München, Germany

Batch number: 10397794

Printed by Printforce, the Netherlands